Old
Electrical
Wiring

About the Author

David E. Shapiro has worked as an electrical contractor and consultant since 1980, and is IAEI- and state-certified as an inspector and plan reviewer. He has authored numerous articles on electrical work and safety, which have appeared in publications such as *The Washington Post*, *Practical Homeowner*, *New England Builder*, *EC&M*, and *IEC Quarterly*, and has a long-running column in *Electrical Contractor* magazine.

Old
Electrical
Wiring
Evaluating, Repairing, and Upgrading Dated Systems

David E. Shapiro

Second Edition

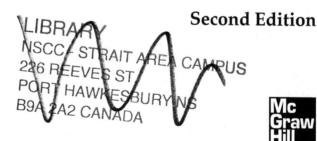

Mc
Graw
Hill

New York   Chicago San Francisco
Lisbon London Madrid Mexico City
Milan New Delhi San Juan
Seoul Singapore Sydney Toronto

The McGraw-Hill Companies

Cataloging-in-Publication Data is on file with the Library of Congress

McGraw-Hill books are available at special quantity discounts to use as premiums and sales promotions, or for use in corporate training programs. To contact a representative please e-mail us at bulksales@mcgraw-hill.com.

Old Electrical Wiring, Second Edition

1 2 3 4 5 6 7 8 9 0 WFR/WFR 1 9 8 7 6 5 4 3 2 1 0

ISBN 978-0-07-166357-1
MHID 0-07-166357-6

The pages within this book were printed on acid-free paper.

Sponsoring Editor
Joy Bramble

Proofreader
Dev Dutt Sharma

Acquisitions Coordinator
Michael Mulcahy

Production Supervisor
Pamela A. Pelton

Editorial Supervisor
David E. Fogarty

Composition
Aptara, Inc.

Project Manager
Indu Jawwad, Aptara Inc.

Art Director, Cover
Jeff Weeks

Copy Editor
Sheila Johnston

Contents

Foreword

While there are many, many "how-to" electrical wiring books available to us, I am not aware of any "How don't" books regarding and describing the present state of our wiring in older residential and commercial buildings. As a working Electrical Inspector, I only see the tip of the iceberg, as many of the violations contained here are abated (or concealed) prior to the inspection. The author's trouble call [bringing him] to the property is not adversarial (as inspections are), but in response to a problem of no power, so he is given full access to all the questionable practices that may be the cause of his visit.

Incidentally, these violations are always blamed on the previous owner.

Much of the older equipment pictured and described date back to the era when a 30-amp service with two-plug fuses was the norm and most of the present-day appliances did not exist. This does not deter the present homeowner when he or she buys a four-slice toaster and an electric coffee pot and plugs them into the one receptacle that was available. When the fuse or circuit breaker opens, then the creativity that is demonstrated in the book is revealed. Chapter 5 describing switching schemes brought back memories of those same schemes that were "cutting edge" 70-plus years ago.

ANDRE CARTAL
Princeton, New Jersey

Foreword to First Edition

The electrical construction industry is so fast moving that it takes a lot of effort on the part of building owners, electricians, electrical contractors, inspectors, apprentices, and others to keep up with new methods, equipment, wiring materials, and tools being introduced to the marketplace in a steady stream, not to mention absorbing the revisions to the *National Electrical Code*, of which there is a new, large batch every three years.

Well, you can relax now, for this book looks back to the way it used to be done, and there is no urgency for you to absorb it all at once. As with most technical books, *Old Electrical Wiring* does not lend itself to being read through in one session. Rather, it is a book you will find to be a valuable reference work. A lot of the information is applicable to your current work: safety practices, customer relations, inspector relations, etc., but keep the book handy for when you run into something you don't recognize. An installation completed last week is "old work," and sometimes such a job can be as poorly done as a really old job, but in general what is covered in *Old Electrical Wiring* is wiring done years ago. The trick is to try to figure out what was in the mind of the electrician at the time the job was done originally. Just because the work is old does not mean it was not well done, nor is it assured that a job done yesterday is necessarily first-rate. *Old Electrical Wiring* will help you make the judgments necessary in sorting out the good from the bad.

David Shapiro has done a fine job of researching the material for this book. The use of electricity in buildings is over 100 years old, so there are no persons alive who have seen it all. You will find that reading *Old Electrical Wiring* is just like talking to an "old timer," a friendly, talkative one at that, and one who speaks your language.

I notice the mention of illegal three-way hookups referred to as the Carter System. Over 50 years ago, when I was an apprentice in San Francisco, we referred to these illegal and unsafe three-way circuits as "Los Angeles three-ways." I later learned that electricians in Los Angeles referred to them as "San Francisco three-ways." In any case, pointing the finger does not accomplish anything. It does not matter who messed it up—it only matters that you know how to fix it, and *Old Electrical Wiring* will help you do that.

Insulation deteriorates with age, so one of your major tasks is knowing when you can salvage something, or whether it must be replaced. This book will help you make those decisions.

Experience is of great value when troubleshooting an old job, but if you don't have the experience yourself, *Old Electrical Wiring* gives you an insight into the experiences of uncounted others to help you. By the way, when troubleshooting, be sure to carefully listen to the owner or occupant who is trying to let you know what happened. Even though you may not speak the same language, technically, if you listen very carefully you will probably get a helpful clue.

I wish this book had been around years ago—it would have been most useful to me, and I know it will be so for you.

W. Creighton Schwan
Hayward, California

Acknowledgments

Many people helped this book come into being by bringing me to where I could write it. Some helped me develop as a writer, some as an electrician, and most as a person. They include the following.

My parents:

> Lillian Shapiro-Michael, an example of integrity and of working to make a positive difference; and
>
> Louis Leonard Shapiro, who showed me integrity, and the pleasure and value to be derived from physical work. He also eased my way to join the IBEW.

Mr. Brown, a grade-school English teacher.

Dr. Leslie Horst. Her coaching made possible my first success as a writer of substantive material—my master's thesis.

Arthur Hesse and Ronald Kotula, chief inspectors whose professionalism, courtesy, knowledge, and good will helped make me a competent master electrician.

Creighton Schwan, my mentor, not the least as a writer for the electrical trade; an urbane, erudite, ever-civil, ever-helpful columnist and code authority.

Various trade magazines which provided learning through their articles and responses and enabled me to get my feet wet writing about the industry.

The editors for whom I wrote columns, including Walt Albro, Nancy Bailey, and Jeanne LaBella; they helped me gain experience and credibility writing about safety.

IBEW Local 3, which gave me a start in the electrical trade, and taught a wonderful attitude of workmanship and pride.

Bill Zinsser, author of *On Writing Well*, who helped me value clear language.

Wilbert J. McKeachie, whose writing gave me some understanding of effective teaching.

My customers, who have trusted me and supported me for so many years and whose electrical systems provided many of the lessons conveyed in this book.

The International Association of Electrical Inspectors, IAEI—a place where persons of good will in this industry come together to further electrical safety.

And finally, my close family, friends and lovers over the years, including most especially Mary Jo Hlavaty, Judith Oarr, and Sheila Wolinsky, who supported me through the twists and turns of my career path with trust, advice, and more.

I offer deep thanks to the following people who contributed more directly and immediately to this book.

M. J. Hlavaty provided major help with copyediting.

Charley Forsberg provided information on PVC.

Dr. Jesse Aronstein reviewed the material on aluminum, early on.

Joseph Tedesco, a man with strong knowledge and strong opinions, reviewed an early version of the entire book and offered feedback.

Sam Levinrad provided exhaustively detailed feedback, giving the author, and readers, the benefit of his very wide knowledge and experience in the electrical industry.

Creighton Schwan provided equally helpful and careful comments. He also provided the entrée to Hal Crawford, which smoothed the way for the making of this book.

Dave Noonan at Bryant offered useful comment on the glossary and other bits.

Arthur Hesse looked through an early version of the text and confirmed the reasonableness of the suggestions in Chapter 12, "Inspection Issues," from the perspective of a universally respected chief inspector with decades of experience and the highest reputation for integrity.

Additional Acknowledgments for the Second Edition

In the first edition, I thanked some of the people who brought me this far as a word-lover, a writer, and an editor, and of those who helped me gain competence in at least one small area of the vast field of electrical work. In addition, some have helped me specifically with preparing this second edition.

Alexander N. Brittain volunteered an intense bout of presubmission editing.

Mary Jo (Hlavaty) Shapiro, my sweetheart of the last 17 years, also is the person who introduced me to the world of Information Architecture, where I discovered the concept behind most of the captions accompanying the illustrations. She also provided the feedback of an experienced editor regarding both the captions and the illustrations themselves, enabling me to improve some and discard others, while maintaining the contract-required number of 300. My loyal wife also shouldered more than her share of domestic chores toward the end of my editing drive.

A good number of loyal customers with old homes continued to rely on me to serve them to the best of my ability during various degrees of unavailability or incapacitation.

Andy Cartal, the warmest IAEI colleague I have had the great luck to become friends with, trusted me with frank feedback on the initial set of illustrations, and then bore down a week before the submission date to go over the text so he could tell you what he thinks of the educational material.

Dr. Jesse Aronstein, researcher and consultant, was kind enough to allow me to include findings from his excellent work evaluating unreliable circuit breakers, including new ones that have not previously seen print.

John Sleights, forensic E.E., was generous with his information on older cables, allowing me to share findings that will appear here just about the time they are printed in a journal.

J. D. Grewell, my most-knowledgeable ASHI colleague, who takes excellent photos, generously permitted me to use six of them, specifically Figs. 2-12, 2-20, 2-21, 10-8, and 10-21.

C. Larry Griffith, electrical worker, inspector, instructor, and IAEI colleague, shared a number of his experiences—and his opinions based on sharp thinking and long experience. Mary Jo and I value our many years of workshop-sharing and friendship with Larry and Bobbi.

Alan Nadon, Chief Electrical Inspector for Elkhart, Indiana, is another good friend and colleague, who helped me understand some of what I read about earlier practices. He started out in this trade heating soldering irons to hand up to his brother, who was standing on a ladder making up joints and splices.

F. Stephen Masek, colleague and superb documenter of his inspection jobs, generously granted permission to reprint his photos of knob-and-tube wiring, as Figs. 3-26, 4-8 and 4-9, and contributed the collection of porcelain electrical products used in Figs. 3-31 to 3-35.

W. Creighton Schwan, my mentor, honored me once by becoming my friend; again by recommending me to take over a book he had edited for decades; a third time by choosing me to be coauthor of his final book, *Behind the Code*, and leaving it to me to bring to print; and finally by asking me to visit him, as he approached death; and finally for bequeathing his electrical library to me, including dozens of old Code books, and all his columns. Those old NECs have allowed me to provide most of the adoption dates you will find in this edition.

The University of California library system did us all a signal service by scanning some of their old, out-of-print and out-of copyright books, and making the images available to us all.

CHAPTER 1

Introduction

This second edition is both shorter (to lower the price) and more focused on practical descriptions of how I deal with old wiring than was the original *Old Electrical Wiring*. Old wiring goes back to the last years of the nineteenth century, even though power wiring was unknown 20 years earlier. This is demonstrated by its omission in Fig. 1-1 and Fig. 1-2, which show a comprehensive engineering handbook whose only electrical portion concerned lightning protection. However, I can't focus on wiring too old for me to have worked on. For example, I've read of the brass device preceding the red hat as an antishort bushing for Type AC armored cable (BX), called a thimble. I've even inspected work by a journeyman who had come across some of these on a job. However, not having dealt with them myself, I don't know whether they require any special care, pose any special hazard, or—in short—have any practical significance except that they would show that the wiring on which I was working was quite dated, and they might provide an unusual keepsake.

It is critical that you don't let yourself be deceived by the practical focus into thinking either edition is a how-to manual, whatever language—in the current edition, or in the original—may seem to suggest otherwise. It is not. I have seen too many jobs botched because inexperienced workers followed written instructions without having someone sufficiently knowledgeable on hand.

I believe strongly in educating laypersons: homeowners, to avoid mistakes such as Fig. 1-3; maintenance people, to insist on not permitting common practices such as the one I show in Fig. 1-4; and even plumbers and appliance installers, not to suggest, or go along with, layouts such as Fig. 1-5. Professional carpenters or cabinet installers undoubtedly chose to destroy the integrity of the cover plates in Fig. 1-6, to avoid calling in an electrician, or even shaving 1/4 in out of the bottoms of the cabinets. However, I don't attempt that sort of education here. I provided an overview for laypersons in my second book, *Your Old Wiring*; and I have other basic books in mind, though not in preparation. It is part of every electrician's and inspector's business to mention items such as clearances—not only those specified in

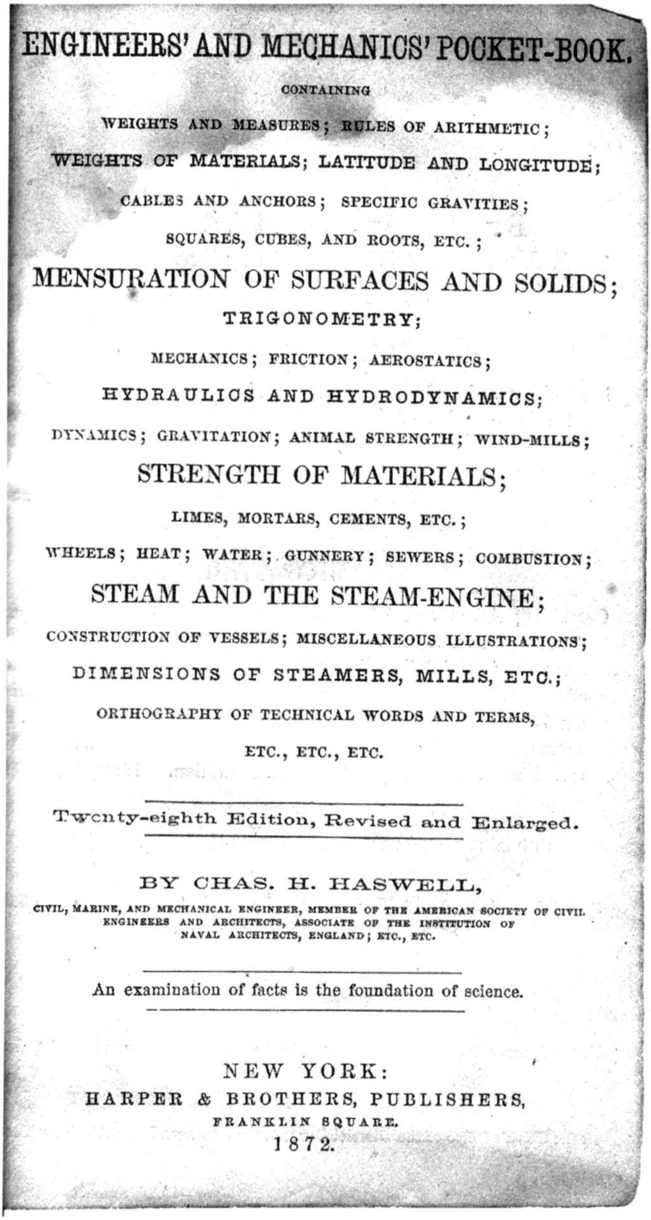

ENGINEERS' AND MECHANICS' POCKET-BOOK.

CONTAINING

WEIGHTS AND MEASURES; RULES OF ARITHMETIC;

WEIGHTS OF MATERIALS; LATITUDE AND LONGITUDE;

CABLES AND ANCHORS; SPECIFIC GRAVITIES;

SQUARES, CUBES, AND ROOTS, ETC.;

MENSURATION OF SURFACES AND SOLIDS;

TRIGONOMETRY;

MECHANICS; FRICTION; AEROSTATICS;

HYDRAULICS AND HYDRODYNAMICS;

DYNAMICS; GRAVITATION; ANIMAL STRENGTH; WIND-MILLS;

STRENGTH OF MATERIALS;

LIMES, MORTARS, CEMENTS, ETC.;

WHEELS; HEAT; WATER; GUNNERY; SEWERS; COMBUSTION;

STEAM AND THE STEAM-ENGINE;

CONSTRUCTION OF VESSELS; MISCELLANEOUS ILLUSTRATIONS;

DIMENSIONS OF STEAMERS, MILLS, ETC.;

ORTHOGRAPHY OF TECHNICAL WORDS AND TERMS,

ETC., ETC., ETC.

Twenty-eighth Edition, Revised and Enlarged.

BY CHAS. H. HASWELL,

CIVIL, MARINE, AND MECHANICAL ENGINEER, MEMBER OF THE AMERICAN SOCIETY OF CIVIL
ENGINEERS AND ARCHITECTS, ASSOCIATE OF THE INSTITUTION OF
NAVAL ARCHITECTS, ENGLAND; ETC., ETC.

An examination of facts is the foundation of science.

NEW YORK:

HARPER & BROTHERS, PUBLISHERS,

FRANKLIN SQUARE.

1872.

FIGURE 1-1   Electrical engineering didn't exist in 1872.

534   LIGHT.

LIGHT.

LIGHT is similar to Heat in many of its qualities, being emitted in the form of rays, and subject to the same laws of reflection.

It is of two kinds, *Natural* and *Artificial;* the one proceeding from the Sun and Stars, the other from heated bodies.

Solids shine in the dark only at a temperature from 600° to 700°, and in daylight at 1000°.

The *Intensity of Light* is inversely as the square of the distance from the luminous body.

The *Velocity of the Light of the Sun* is 192500 miles per second.

The Standard of Intensity or of comparison of light between different methods of Illumination is a Sperm Candle "short 6," burning 120 grains per hour.

Loss of Light by Use of Shades.—(F. H. Storer.)

Glass, etc.	Thick-ness.	Loss.	Glass, etc.	Thick-ness.	Loss.
	Ins.	Per Cent.		Ins.	Per Cent.
American enameled	3⁄16	51.23	Window, double, Eng. ...	1⁄5	9.39
Crown..................	1⁄5	13.08	" " German.	1⁄2	13.
Crystal plate............	1⁄2	8.61	" single, " ..	1⁄16	4.27
English "	1⁄3	6.15	" " ground..	1⁄2	65.75
Porcelain transparency ..	3⁄16	97.68	" green	3⁄16	81.95

ILLUMINATION.—GAS, LAMPS, AND CANDLES.

Comparison of several Varieties of Lamps, Fluids, and Candles with Coal* Gas, deduced from Reports of Com. of Franklin Institute, and of A. Frye, M.D., etc., etc.

Lamp and Fluid.	Intensity of Light.	Ratio of Cost per Hour.	Light at Equal Cost.	Time of Burning 1 Pint of Oil.	Relative Costs for Equal Lights.
				Hours.	$ Cts.
Camphene	1.75	.57	3.08	9.31	.32
Carcel, Sperm oil, *maximum*	2.15	1.22	1.8	6.32	.56
" " *mean*	1.22	.86	1.35	9.87	.74
" " *minimum*	.69	.67	1.2	14.6	.83
" Lard oil	.77	.76	.97	11.3	1.03
Gas	1.	1.	1.	—	1.
Semi-solar, Sperm oil	1.15	1.25	.93	6.75	1.07
Solar "	1.76	1.09	1.55	8.42	64

Candle.	Burns.	Intensity of Light.†	Light at Equal Costs.	Cost with Equal Light.	Cost compared with Gas for Equal Light.
	Hours.				
Diaphane	6.6	.7	.5	2.08	15.1
Palm oil..................	6.6	.7	.77	1.32	10.5
Spermaceti, short 6's............	8.	.8	.54	2.16	16.2
Tallow, short 6's, single wick....	6.	.58	.85	1.	7.5
" " double "	5.5	1.	1.	1.46	7.1
Wax, short 6's	9.	.8	.61	1.96	14.4
" long 4's................	13.	—	—	—	—

* City of Philadelphia.
† Compared with a fish-tail jet of Edinburgh gas, containing 12 per cent. of condensable matter and consuming 1 cubic foot per hour.

FIGURE 1-2 Lights burned gas, oil, or wax.

FIGURE 1-3   Loose cables are ripe for misuse.

National Electrical Code (NEC) Art. 110, but also utilities' clearances like those violated by neglected bush-trimming in Fig. 1-7.

Rather than providing directions or focusing on the quaint, I describe my experiences with old wiring; that is what my title means to convey. I look at systems that mix older and newer wiring, because there are lots of those. Unsurprisingly, many systems whose wiring is not necessarily antique have been screwed up over the years due to confusion or, more commonly, ignorance, or at least someone

FIGURE 1-4   Electric closets are frequently misused for storage.

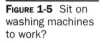

FIGURE 1-5 Sit on washing machines to work?

FIGURE 1-6   Damaged cover plates don't fully cover.

FIGURE 1-7   The service point mustn't become inaccessible.

not understanding what they were dealing with quite well enough. Figure 1-8 shows an example of ignorance, resulting in work that was simply kludged; whoever added the 12 American Wire Gage (AWG) branch circuit showed no acquaintance with concepts of ampacity and overfusing, among other issues; all they cared about was finding breakers providing 240 volts. This sort of problem also earns some notice, because I haven't found other texts focusing on the out-in-the-field reality. To save space, this edition is more tightly focused on genuinely old wiring than was the first.

While this edition is less chatty than the first, I know how important examples are. I replaced unclear illustrations from the first edition and more than tripled their number, at the expense of text, for contract reasons. My descriptions assume that readers are professionals, with the training, experience, and resources this implies. Daniel Friedman's Web site for home inspectors (http://www.inspectapedia.com/electric/electric.htm), which contains a fine section of illustrated alerts to hazardous electrical products, does the same.

One consequence of focusing on professionals is that I use jargon more than in the first edition; however, I do attempt to define or at least expand most abbreviations at their first appearance and again in the

FIGURE 1-8
Overfusing often goes with doubling up on overcurrent protection (OC).

revised glossary. Another is that I give much less space to explanations of why something is important. For example, Fig. 1-9 may look irregular to someone other than a trained electrician, but I hope my readers recognize how serious it can be for a meterbase to lose structural support and hang from cables, as in this picture.

I welcome readers who lack the experience this presumes, but I have to caution you: my descriptions may seem overly telegraphic, despite the glossary in the Appendix following Chapter 11. You need a certain familiarity with the NEC and product listing requirements, and in addition the skills and related background of someone experienced. If that's not your situation, do get someone who knows the ropes to help you. Please don't try to extrapolate the descriptions without this, because I've seen the results of extrapolation in installations such as Fig. 1-10, in which a professional can see the vast difference between the workmanship—and legality—of the old wiring and the new. Closer mimicking *still* would not have resulted in something legal.

Figure 1-9
Utility-side parts
can't be poorly
secured.

Figure 1-10 New
work rarely models
carefully on old
workmanship.

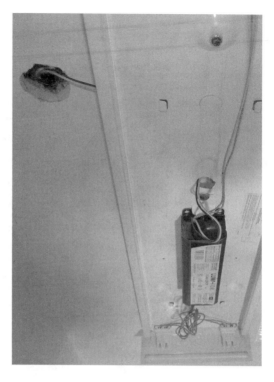

I sympathize with people who craved actual instruction when they read the first edition of *Old Electrical Wiring*, or when they read the *Code*, or even Underwriters Laboratories, Inc. (UL)'s *White Book*. But none of these, including this edition, is the place to learn about why Fig. 1-11 shows at least four violations, or how to add the luminaire legally. It comes down to this: if you hope to work on atypical wiring, I know of only two ways to learn halfway safely. One is to have an old-timer to supervise you, and I mean an old-timer who has a high respect for safe practices. How old? They don't need to be old enough to have worked on electric vehicles from the early twentieth century, using the book in Fig. 1-12. That's not what you will be working on, so it's irrelevant to the Article 100 definition of "Qualified Person." The other is to be *considerably* experienced at troubleshooting, and to approach this work with ever so much caution, alertness, and awareness of how much can be waiting to bite you. After all, many of the practices that were used nearly a century ago are still around— not to mention some of the materials, as the 1917 advertisements in Fig. 1-13 and Fig. 1-14 show. However, while some extremely dated items such as those you see in Fig. 1-15 still exist and may even do

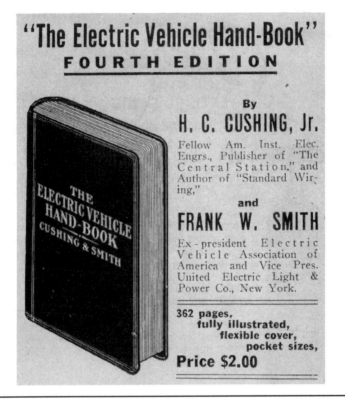

FIGURE 1-12 Electric vehicles go way back.

their jobs, these designs are too far behind our present standard not to warrant our completely, or at least largely, replacing them when re-doing these portions of a wiring system. Some vestiges, such as rigid metal conduit (RMC) with couplings that were sealed with white- or red-lead paint, carry danger for us when we work on them—by virtue of their high workmanship standard!

Because I can't be there to look over your shoulder, this book is not designed to be used as a training manual. Anyone who reads it as instruction is misapplying it. It also is not as wide-ranging as the first edition in looking at old wiring and the issues associated with it. In part, this is to make the new edition more affordable. If you can't get hold of the first edition, or if you want something even more compre-hensive, the *American Electrician's Handbook* may at least list the items you're wondering about, although without explanation or discussion. There are reasons that book has stayed in print since 1913. Sometimes you might wonder whether the installation in front of you ever was legal, as it is clearly in violation of the Code you are used to working

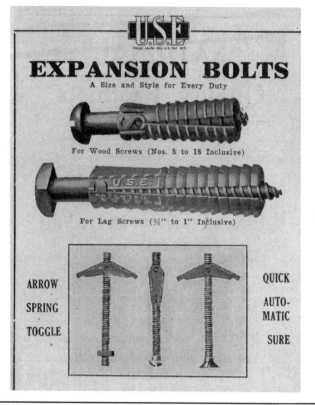

FIGURE 1-13   Some hardware we use goes back a century.

with. The book you're reading may offer some limited assurance that the workmanship wasn't haphazard. *Behind the Code,* by Schwan and Shapiro, offers a framework for understanding the Code's changes and provides a sampling of the answers you may be seeking. The basic process is described at http://www.nfpa.org/categoryList.asp? categoryID=162&URL=Codes%20&%20Standards/Code%20development%20process/How%20codes%20and%20standards%20are%20developed. However, *Behind the Code* talks about how a number of the rules we see how come to be, which can help you—or an apprentice or a customer—to understand those *Code* requirements, and to make your own decision about respecting them.

To my loyal readers and colleagues from *Electrical Contractor* and *IAEI News,* welcome. To those who bought the first edition and have been waiting patiently for an update, I hope you enjoy this, and accept the tradeoffs that were necessary to get it back into print.

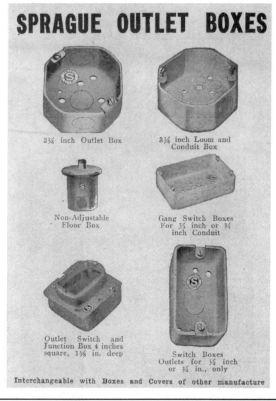

FIGURE 1-14 Metal boxes have changed a little.

FIGURE 1-15 I don't reuse old grounding electrode conductor (GEC) clamps.

Overview

Chapter 2 is where I begin to talk about how I approach the very biggest and most important issue I deal with most days as I approach old wiring and also how my approach has shifted over time. In the first edition, I dove right into how I judge, fix, and replace old wiring, leaving safety at the back. Well, I was a dozen years younger, and had not yet recognized some prices I was in the process of paying.

We have the right to spend the necessary time to keep ourselves safe. Even electricians who feel the need to put this out of mind during the effort to finish a job deserve to start with the facts. This said, while the work is risky, if all I talked about were negatives I'd be dishonest, or at least a fool. I close this chapter by acknowledging and underscoring the value of this work. In a sense, the entire book serves that purpose.

Chapter 3, Troubleshooting, is where I wade into explaining how I work. Once we're there, looking at how to accomplish the customers' agendas, the subjects that I cover in Chapter 3 could plausibly fit in many other chapters. Another way of looking at the design is that troubleshooting is my subject for the greater part of this book.

Where do I know we have a good circuit? What's the first place I don't? What's most likely to have gone wrong? Where's the best place to look for it? I don't necessarily give the same answer to these last two. Then I start talking about choosing the appropriate fix and determining what else needs addressing besides the problem I was presented with.

Chapter 4 goes a little deeper into certain issues, such as cleaning up wiring that was quite illegal. It may have been that way from the start, or it may have turned into a mess after someone tacked on extensions they considered good enough simply because they worked, or after nonelectrical work made some aspect of the electrical system unsafe or unusable. The last part of this chapter starts to look at grandfathering. I find the common story much as Andy Cartal noted in the Foreword: we're almost always told, "It was that way before I came."

Chapter 5 is very much a continuation of Chapter 4. While I talk about various switching issues, I focus on one outdated system, a design that is never safe or legal in power wiring, which I know as the Carter System. Some old switching systems can be mighty baffling without knowing about this. I talk about how I find Carter System switching, how I sniff out what only looks like it, and how I fix it.

Chapter 6 moves on to other puzzles, pitfalls, and hazards such as outdated designs that may still be present. These range from what's legitimately grandfathered, such as old cooking circuits and outlets, to a related violation—the service upgrade that converts a grandfathered circuit to an illegal downstream ground—to wiring that no longer is in the *Code* or that lingers but is rarely, rarely used. I also review some issues concerning aluminum wiring, especially branch circuit wiring. I skim lightly past certain low-voltage systems that tend not to justify patching up.

Chapter 7 zeroes in on issues that I have found more frequently, or only, on commercial and other nonresidential jobs. I also touch on electrical designs that are sufficiently interesting or important to bring up but that are so specialized that I have no personal experience with them. Then I move on to subjects such as motor work and the types of load that can develop when a commercial operation grows. The last challenge that I describe is the institution that acts as though it is exempt from the rules.

Chapter 8 steps back in to the typical "can you do this for me?" job in an old building. Retrofitting paddle fans is a common task in houses lacking central air conditioning or modern circuiting. The job sounds simple, it can be simple, but very often in older buildings it is not simple to accomplish this legally and safely. I have spent the last few decades figuring out viable ways to tackle jobs bearing just these characteristics.

Then I branch out, providing more of the "whatever is that?" information I began sharing in Chapter 6. I explain what it can mean when a long-functioning switch mysteriously stops working when tipped out of its enclosure and functions again when put back—without anything being damaged, loose, or out of specification. There are antique designs that are never found in stock at the distributor, but that remain legal, and can be ordered—at a premium—if the customer wants them replaced in kind. More of those that I describe are, however, long gone from suppliers' shelves, deservedly.

Chapter 9 follows naturally. How do I decide that what's in place is in too bad a shape to let live? What are some of the ways I go about upgrading it?

First come the best approaches, true fresh starts. After these, I move on quickly to tricks and finesses that can save time, mess, and expense, or at least measure out these difficult aspects in more-tolerable doses. They do represent compromises, but old work demands a judicious dose of realism.

Realism does not mean shutting one's eyes to *Code* and safety issues, and at no time is this challenge more vivid than when inspecting and consulting. Chapter 10 focuses on one of the most critical of all topics in this book, second in order of importance only to Chapter 2's discussion of safety. The reason that I placed it so far back is that to inspect work competently a person has to understand how it is done. The words in the *Code* book, handbooks, and product specifications have little meaning outside the contexts in which they are applied.

I talk Here, about some of the problems faced in inspection and in dealing with inspectors. This chapter covers both jurisdictional inspections and private consultations. There is nothing that keeps up morale and the quality and safety of the wiring in an area like a competent and honest inspection department that is given the necessary resources to perform its job meticulously, working with the rest of the licensed, educated electrical community. Even when this falls short in some way, though, inspection remains the responsibility of every one of us who takes on the responsibility of an electrical job. I hope this chapter conveys that message.

Some of the snarliest threads that we are called on to unravel in old work are those involving genuine antiques that beg to be respected as the carryovers they are. Chapter 11 wraps up this introduction to old wiring. First, I return to the complicated judgment involved in grandfathering and discuss how I perform research to help me date an installation. Then I move on to the special needs of buildings whose owners want to preserve them as intact as possible because of their age or their place in history—and the special *Code* status that jurisdictions sometimes grant to the systems in these buildings.

After Chapter 11 comes a glossary in which I provide a few definitions to help colleagues from different areas. In addition, I translate the abbreviations and colloquialisms that are familiar to all electricians, for the benefit of novices who are reading this to pick up what they can. I don't explain all the jargon, but I do expand many of the terms, so that you can find definitions by referring to the NEC.

CHAPTER 2

Hazards and Benefits

U nderstanding the material in this chapter is a top priority. Old work carries many subtle dangers that took me years to recognize. Only by paying painful, unnecessary prices did I learn to grant these dangers more of the respect they demand.

Although the risks overlap with those found in any electrical work, in this chapter, I focus on the hazards to which I have exposed myself in working on old wiring and in old buildings. I close on a positive note: a look at how understanding old wiring has benefited me as an electrician, consultant, and inspector, whatever the job.

Dangers of Old Wiring

I didn't start out equipped for old work. What I was taught as an apprentice, helper, and journeyman in dealing with new installations only taught me the beginnings of how to troubleshoot, evaluate, and repair old systems safely. I have found that working on, consulting on, and even inspecting work on old wiring, and generally working on existing residential or small commercial systems, carry their own risks. For one thing, modern systems are more likely to have been installed and maintained properly. However, short-sighted managements do allow even some industrial and large institutional systems to suffer neglect, unnecessary deterioration, and incompetent modifications. Not only is this physically dangerous, not only does it make the property and its users less safe, but it makes work on the system more dangerous to me. Insisting on safe work, against this background, has lost me customers. That once-in-a-while loss is a price I have learned to pay.

Partly because old buildings are less predictable structurally, old systems are more likely to have undergone undocumented changes. Also, in old work I am even more frequently expected to work around clutter, which can magnify those same hazards.

Special Risks in Old Work

National Fire Protection Association (NFPA) 70E, the *Standard for Electrical Safety in the Workplace*, arrived in 1979. While it was NFPA's best seller after the *National Electrical Code* (NEC), it is more a statement of principles than a list of specifics. It has value, but doesn't take me as far as I need.

Much of my work has been "old work," whether I'm there for a minor repair, renovation, inspection, or rehabilitation. Performing maintenance and repair of already-wired structures inherently means I'm not first. I find this makes it more risky than most new construction of similar AIC and other relevant ratings in a number of ways. To begin with, systems that I have not installed and maintained lack a design philosophy I know I can rely on. I respond accordingly.

Example   Just as I normally test my tester before using it to test and—usually—again after getting a "safe" reading, I have learned to double-check the accuracy of *every* circuit directory—including that of my home, even though I personally rewired it in the early 1980s, creating a rational layout—rather than assuming I know how to shut off power to any point.

When I come on a job that has been worked on by others, there is some reason to fear that what has been done previously may not be as it should. I may wonder why the customer has changed electricians. There are many possibilities. Some carry physical danger to me, or at least potential liability, or imply extra hassles.

Problems Added by Older Wiring

Old buildings pose far, far greater threats than relatively new ones for four reasons.

- First, old systems are likely to show some deterioration. I have never had a customer give me the nod to megger their wiring "just to make sure," and yet even meggering wouldn't test whether I and they were safe from anything but insulation weakness that was ready to show itself at that moment.

- Second, the longer a system has been around, the greater the chance that along the way someone incompetent (perhaps from ignorance) or careless worked on it. Commonly I find layer on layer of very variable handiwork, as in Fig. 2-1, a 4 in-square (1900) box installed with nonmetallic-sheathed (Type NM) cable, then probably with BX added, and newer NM alongside the old. It shows short wires, unstripped sheath, and illegal splices. Like Fig. 2-2, its grounding/bonding is clearly homeowner/handyman work. I see that

FIGURE 2-1 NM added often shows poor grounding.

less frequently inside panels, as in Fig. 2-3, but I can't ignore it anywhere. Figure 2-4 shows another familiar result of do-it-yourself (D-I-Y) ignorance: old cable unsecured to the nearby 2-by-4, and unsecured to the new enclosure, doubly threatening the connections inside. Figure 2-5 shows a conductor with thermoplastic-heat-wet location insulation (THW) that had its box removed in the course of other abuse,

FIGURE 2-2 NM added often shows poor bonding.

Figure 2-3
Clueless splicing
and open KOs
show ignorance.

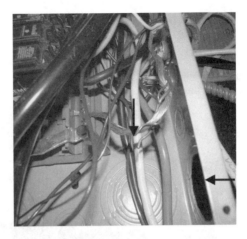

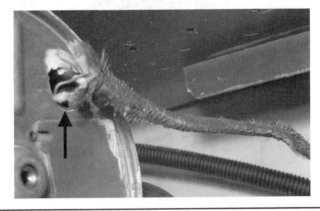

Figure 2-4 NM lacks connector and support–near wood.

Figure 2-5 Heat
that kills rag wiring
even damages a
conductor whose
insulation is rated
to withstand
moisture and heat
(THW).

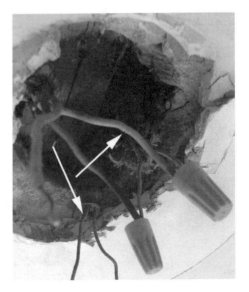

Figure 2-6 New work boxes often are misapplied.

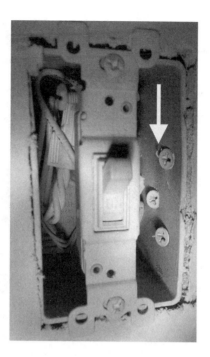

its insulation aged so it cracked over a few decades. Figure 2-6 shows ignorance of a less-obvious rule, the requirement for adequate box support, exposing insulation to an unnecessary risk; Fig. 2-7 illustrates another "minor" rule break, where an installer disregarded the need to secure cables to eliminate strain. Figure 2-8 shows an amusing wrong guess, amusing except for the fact that the protection created by complete enclosure has gone out the window. Figure 2-9 shows another omission of complete enclosure, except that the workmanship is sloppy. It could have been a tired electrician's work, but something about it tells me it was amateur.

- Third, electrical doctrine and standards have changed markedly over the years.

Example The following problems may or may not be widespread; I have not run into them, but my mentor Creighton Schwan did. Hazards have been created as the result of ground currents flowing through the wire mesh under stucco, due to one old doctrine. In the early 1950s, in the San Francisco Bay area, boxes fed by two-wire NM were grounded by running a bare conductor on the face of the studs and wrapping it around the box-mounting nails. These high-resistance grounds caused problems when shorts through them energized the wire-mesh plaster lath.

FIGURE 2-7
Unsecured cables
often show strain.

A slightly earlier practice was worse: the World War II-related copper shortage caused installers to use iron wire in this way. In 1989, a Pennsylvania contractor wrote about a similar problem that occurred in his town, which resulted in energized window frames when an ungrounded outlet ground-faulted to the box, from there to the box

FIGURE 2-8 Good
intentions don't
make up for
ignorance.

FIGURE 2-9 'Any old cover' isn't good enough.

cover, and from the ground wire to the wire lath. This resulted in energized window frames.

- Fourth, even when they were state-of-the-art, most of those methods and materials that have passed out of the NEC, or have been modified substantially, were more dangerous than the modern ones that have replaced them. I'm not saying that NM run today to a self-holding plastic box is safer than the fine old approach of using pipe-and-wire to feed a 1900 nailed to a stud, or that a self-holding plastic box in drywall is more secure than a metal box screwed to wood lath. However, today's grounded NM fed into a plastic box is safer than ungrounded NM 75 years ago feeding into an often-crowded Bakelite box, or a small metal box. Those antiquated systems are yet more dangerous now, deterioration aside, simply because they contain elements that today's workers find unfamiliar and perhaps totally unexpected. I certainly still run into surprises. Much of this book is devoted to describing experience that may help you minimize this in your own work.

A moment back, when talking about "old work," I mentioned the fact that taking responsibility for electrical work carries an additional layer of risk beyond the hands-on physical danger associated with actually performing the work. These dangers are magnified in work on old wiring systems.

The first edition looked more closely at the risk of innocently opening someone else's can of worms only to have a hungry lawyer peck
my eyes out. For now, simply consider the fact that as a contractor I
offer assurances, implicit or explicit, in four directions.

- First, there are unknowns associated with old wiring repairs
  that can cause problems for my customers. As the point man,
  I may be reassuring people or I may be bearing bad tidings.

- Second, there are questions relating to permits, involving issues of repair versus alteration and of grandfathering.

- Third, I have some responsibility for the safety not only of
  myself but also of others on the job such as inspectors and
  other trades, including those who come onto the premises
  after I am gone.

- Fourth and most especially, I need to leave the site as safe as
  possible for the next electrician, even if it was full of danger
  when I arrived. Unfortunately, I have to decide what's "as safe
  as possible" within the context of what the customer is willing
  to have me do.

Someone who knows nothing of construction work may, in some
ways, be a better risk than someone who thinks he or she knows what
to expect. It's too easy to assume that turning off a light by flicking the
switch, or even unscrewing a fuse, means the outlet is dead. I have
seen plumbers doing this and proceeding to cut adjacent copper water
lines—without testing!

For this reason, I will pass on three stories from a good friend,
long-time Indiana inspector Alan Nadon, for you in turn to pass on
where they will do some good.

Examples   The service to a computer repair shop suffered an open
neutral, and the water pipes were carrying the unbalanced load to
the utility via the ground. Plumbers were working in the street to fix
a leak, and a helper standing in water was hit by that current as he
started to cut into the pipe.

A water softener installer was far luckier. His installation interrupted the metal piping with plastic, and he had just found it amusing
that he could draw an arc between the cut ends. The home insurance
company made him pay for his customer's electronic equipment, her
refrigerator, and her rewire.

A local water utility worker talked about one house where they
could see the water boiling in the meter. His crews *always* use jumper
cables when removing water meters.

Figure 2-10 shows a sump pump that could easily have been
installed by a plumber—or not. Cocky homeowners, unfortunately,

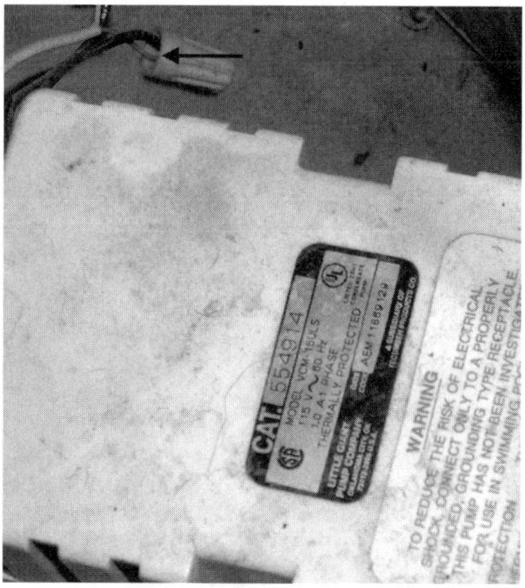

Figure 2-10
UnListed
hard-wiring often
has other
problems.

often choose to take on pump replacement themselves, without respecting their limits. I do try to teach timid customers how to feel more in charge of their premises, but there also are people I eventually decide not to work for, simply because they are so certain they know what's needed, and what's safe—even though I bear responsibility!

In the end, I cannot control what others do, but I need to make it hard for the ignorant to bypass safety measures. This is why, for my own protection, not only tagout but also lockout is desired. This said, especially in a single-family residence (SFR), alone or with an occupant I trust, I feel safe relying on tagout alone.

Example   In rewiring one house (a gut job), I provided at least one ground-fault circuit interrupter (GFCI) protected duplex receptacle at each level, fed by a 20-amp circuit. After labeling the panelboards clearly, I left these circuits on, but turned off all other circuits and tagged them. I pointed this out to the general contractor (GC's) super, and asked him to explain to his drywallers about the safe circuits I had made available for their lights and other tools. Nonetheless, when I came back a day or two later, every circuit in the house had been turned on. I explained the safety issue again to the super, and also to the owner, so they would emphasize it to the builder. This didn't help. Fortunately, no one got shocked. If one of the workers had been hurt, I would have felt bad, and perhaps have been charged with liability. This is true even though I had tried to provide for their safety, and even though ultimately I might not have been held liable.

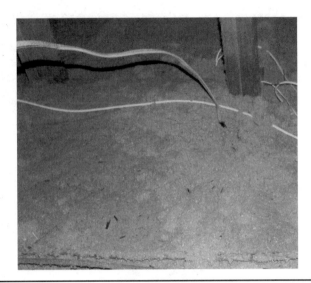

FIGURE 2-11   Many attics contain rock wool or asbestos.

Other Dangers of Work on Older Structures

Dangerous assumptions extend beyond the electrical realm. If I were unfamiliar with loose-fill asbestos such as I may have found in Fig. 2-11, I could have scarred my lungs—as my dad scarred his—as a result of assuming that thermal insulation was going to be a mild irritant rather than a serious carcinogen. (I still don't know whether my substantial fiberglass exposure has caused damage.) The Environmental Protection Agency (EPA) has new rules requiring training so we know what to do when we disturb substantial areas of walls or ceilings that *might* bear lead-based paint, based on their age (http://www.epa.gov/lead/pubs/toolkits.htm.). I also could put not only myself but potentially many others at risk from gas or hot water or steam by cutting or drilling into a gas or other line.

Old work generally is not unique in its safety requirements, but it includes some pretty damaging environments and demands. For instance, access to the workspace often is more restricted. I wish I had learned years earlier from all the specialties that could have saved me eventual grief: audiology, industrial hygiene, and ergonomics are a few. One important lesson that I have learned to apply increasingly is that I don't have to accept what I'm presented with.

Major lesson
For years, I did not bother finding real experts to teach me what protection I do need in various environments, or how to select and apply it. Frankly, I didn't realize the need. Being very fit, I seemed to be doing

fine, and didn't yet realize that I would pay for my ignorance. This was a major personal discovery: when it comes to safety, I need to get information from experts, rather than rely on the fact that I seem to be handling things okay, or "well enough."

Ignorance can be deadly, and even when its results are not deadly it still can be cruelly destructive. I learn of new issues every time I attend a technical seminar. Those of the Association for Preservation Technology are one example outside the electrical industry. In doing so, I continue to discover areas related to job safety in which I didn't even realize I was ignorant.

Example   I once presumed that the main danger from lead was from paint flaking, and little kids somehow glomming onto the paint chips as something to put in their mouths. The source of their risk is more subtle than that. I also once thought the hazard from asbestos was largely from loose fill, and beyond that I probably would have agreed that shredding pipe wrap is dangerous. I know a bit more now.

Example   To save space, I am not going to go into any specific of what one should or shouldn't do—with one exception. I specifically remember someone asking me, "Doesn't that hurt?" as I landed on my knees. "No," I said truthfully, and they walked away, apparently impressed. The cumulative effect was permanent damage. Now, many years later, they hurt, just where I wore away their cartilage and inner padding without realizing it at the time. I am committed to warning people when I see them endangering themselves; I don't take the chance of relying on hints.

Examples   I carry other consequences in my body. I have lost some lung capacity due to cockiness about being able to handle whatever I was breathing; I didn't know this until I took lead abatement training, and for the first time was taught how to choose the right respirator, and was fitted correctly. Of course, filtering does little good when I or the painter or plumber is generating nasty fumes. I have taken chances with fumes, too, and this could be part of what caused my "early stage lung disease"—which in itself has real consequences. I have lost some hearing, and some joint use, apparently as results of not always using suitable protection. One of the job site consequences of the hearing loss is that I can't rely on tapping these days to detect the edges of concealed studs and joists.

Resources

The government offers a considerable amount of guidance in many of these areas. For instance, see http://www.osha.gov/SLTC/etools/respiratory/index.html. It talks about how to pick appropriate protection, when to change cartridges in non-disposable breathers, and more. Another, covering respirator recalls, is at http://www.cdc.gov/niosh/npptl/usernotices/default.html. These are a tiny sample.

I hardly need a Web site to warn me about some of the other hazards. While there may rarely be anyone working around or above me on some jobs, this doesn't eliminate the need for protection of head, hands, and knees from trash. Here, is a partial list of what I've encountered in old houses:

Dust, including lead paint dust

Plaster in various forms including powder, chunks, and slabs

Sawdust

Termite castings

Cockroach castings and eggs

Mouse feces

Rockwool

Asbestos

Old newspaper

Crumbled wire insulation

Nails

Tools

Former food

Sand

Stripped cable armor

Dead light bulbs

Broken glass

Pens and pencils

Scraps of sheet metal

Used tissues

Pieces of brick

My friend Andy Cartal added "bat droppings," and noted, "These can really burn your nose."

Rational Response to These Concerns

How should I have responded these worksite hazards? When the job is large enough, if any of these concerns are applicable, my general contractor needs to consult an industrial hygienist to determine abatement measures and then implement them. If I am not satisfied that the tests have been made and that any necessary protection has been provided, I could try to talk to my local EPA or Occupational Safety and Health Administration (OSHA) people. Well, presumably. In recent years their funding and staffing were so poor that I believe I would have had to make a very convincing case in order to get them to help.

Most of my experience, however, has been with small jobs where the burden is not on a GC. Ultimately, it's my body that is protected or that pays. If I think I could be damaged, I can walk, or I can use a workaround. Sometimes, though, I will compromise with my preferred approach. My body absorbs the cost—and will show it later.

Example   I was inspecting a cable run by a homeowner, a man I knew for a meticulous worker, in a semiaccessible space. Following it, as I preferred, would have meant bruising bits of my body that already hurt. Instead, I shined an adequate light for long enough to confirm that cable appeared to be intact, supported properly, and unspliced. In other contexts, the very same workaround would not have satisfied me.

Many times, those responsible for the job environment are not willing or able to make working conditions safe. Still, if a wall is too dangerous to break into, or doing so would be too expensive, or access would mean an exposure for which I don't have personal protective equipment (PPE), sometimes I will propose alternatives. Power can be run the long way around or on the surface. Wireless or powerline-carrier communications systems can serve in place of low-voltage lines or switch legs.

> **Business consideration**
> Ideally, I would put any significant proposal in writing, and get the customer to sign off, with a witness's signature, some lawyer language up front, and probably a few other safeguards. This, however, is the way my very small business has worked: I ask all my customers to read a general form providing my Terms of Engagement. As I work, I try to keep them updated about what I'm finding and planning. This by itself has worked fine in most cases. I do continue to learn about how to improve my communication with customers.

The Benefits of Learning about Old Wiring

Has all this complication and danger scared me away yet? Clearly not. For perspective, an intelligent motorcyclist both recognizes the risks of bikes—death, injury, and gradual hearing loss are three—and appreciates the benefits. He will tell anyone: "Ride scared."

I act on that same principle as I tackle old wiring: I "ride scared"; it just makes sense. But consider this factor. Like almost any electrician who doesn't stick strictly to new construction, I would have some occasion to deal with old wiring. This was true well before I started to make such work a specialty. It is always good to know more of what

FIGURE 2-12
(Illegal) Old work
may be performed
after a full rewire is
inspected, yet
before move-in.

PHOTO BY
JD
GREWELL

one is up against. In the meantime, it is better for everyone concerned if some people specialize in such work.

It is also true that much of my experience with old wiring applies better than I would have thought to jobs where I'm troubleshooting or evaluating relatively new wiring. That is especially true of wiring that lots of people have worked on. Senior ASHI member J. D. Grewell found the violation shown in Fig. 2-12 in a brand-new condominium that had not even been occupied. He reports that the purchaser wanted an additional wall installed, and the electrical inspector would not pass the job without a receptacle being added in it to comply with the spacing requirements in NEC Sec. 210.52. The inspector, naturally, was not there when it was installed. The only reason Grewell discovered the violation was that drywall had been cut back to help Grewell find the source of a water-leak.

People who know how to deal with old systems will always be needed, to supply consultation, repair, and inspection. This will continue to be true as the definition of "old" shifts with time. This work is a reliable, albeit low-profile niche. Old systems survive amazingly long. The longer I am in this business, the less apt I will be to say of any type of wiring, "Oh, I don't think I'll run across any more of that outside of some museum display." I'm being called in to work on nineteenth-century buildings, now, in the twenty-first century.

Finally, knowing this work has improved my judgment as I install new wiring.

Examples   A "damp location" is one that is subject to moisture but not to thorough wetting or to immersion. Do bathrooms count? A strict construction, one followed by most inspectors, says that a shower stall

Figure 2-13 The actual photo shows copper corroded green.

below the shower head is a wet location, and the area above a wall-mounted shower head may be considered a damp location. The rest of a bathroom, as opposed to a steam room, usually is considered a dry location.

This is reasonable, and dry-location fixtures and hardware installed in bathrooms generally work fine. Having worked in older bathrooms, though, including some that are not wonderfully well ventilated, I have dealt with rust, frozen screws, peeling finish, and damaged lamp sockets. This experience has made me think differently about the "dry" classification as I choose the materials to wire new bathrooms. I am more cautious generally about moisture, which also caused electrical equipment damage in the environment of Fig. 2-13, above a kitchen backsplash, and others shown through this book. Outside, I emphasize caulking, especially around canopies set against brick like the one in Fig. 2-14, but not only them.

I have become a fanatic about writing circuit labels—specific wording, but in lasting terms. Paul's bedroom may become Sally's and then become the office or the storeroom, with or without any carpentry involved. But it is likely to remain the small first-floor room closest to the meter. Unfortunately, the customer sometimes objects to my spending the time to prepare a good directory. This is particularly ironic in view of the fact that I have served some customers for many years without such authorization. I've been coming back again and again to one house to squint at the crowded, inaccurate directories of their multiple loadcenters, onto which I've squeezed what current information I can.

Other violations I have learned to deemphasize. I would not support cable as in Fig. 2-15, or green-tag an installation that included such work. However, in existing wiring, correcting amateurish attempts such as it shows would be far from my highest priority.

Figure 2-14 Most outdoor canopies need caulking.

Figure 2-15 Tape does not approach legitimate support.

FIGURE **2-16** Wood standoffs don't eliminate rust.

Returning to the issue of dampness, experience with samples like Fig. 2-16 taught me how hard it can be to block moisture. Mounting a box or panel on wood rather than against the wall does not prevent rust—and if the outside rusts, connections tend to corrode. However, while Fig. 2-17 shows a considerably rusted old enameled fused disconnect, not only had the inner schematic, Fig. 2-18, survived quite legibly if imperfectly but the fuseholders, conductors, and terminals looked good (Fig. 2-19).

There are always new lessons. As I was completing the preparation of this edition, J. D. Grewell offered me the photos of the installation shown in Fig. 2-20. All these disconnects remained in use, despite the fact that one cabinet was rusted through to show the busbars, shown in close-up in Fig. 2-21. Initially, I was surprised at the indication that sheer age rusted National Electrical Manufacturers Association (NEMA) 3R cabinets outdoors of office townhouses. Grewell proposed an explanation, giving me something new to keep in mind. The wall on which they are mounted is a retaining wall. On the other side, above the disconnects, is a sidewalk, which the building management may treat heavily with snow-melting salt. To resist chemicals over the

FIGURE 2-17 Outside rust may be okay.

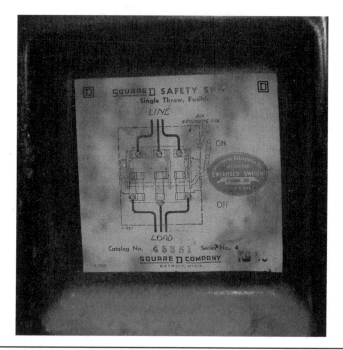

FIGURE 2-18 Even the old schematic can be read.

FIGURE 2-19 The porcelain fuse blocks can withstand modest moisture fine.

years might require not NEMA 3, but enclosures with "P" or "X" in their designations.

Experience with older buildings has taught me to keep my eyes peeled for certain problem children that show up much more frequently than this last poster child. One example is the shop light. Where I notice one, I want to check its source of power. Is it fed by a draped or ty-rapped extension cord? Has it been converted,

PHOTO BY
JD GREWELL

FIGURE 2-20 NEMA 3R is not always enough outdoors.

Figure 2-21 3R or no, the box rusted through.

inappropriately and ineptly, to hard-wiring as was the one in Fig. 2-22? Another, not very different from the first, is shown in Fig. 2-23. In that second case, it may have been installed to save energy, and nobody realized that they were creating a hazard by bypassing the incandescent that had served the location. I've grown familiar

Figure 2-22 A shop light spoiled legal wiring.

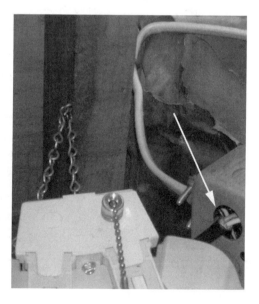

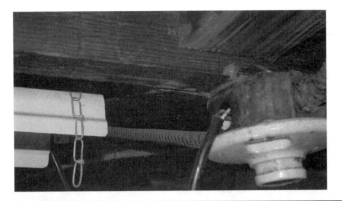

Figure 2-23 The unit lampholder (porcelain) was not satisfactory.

with problems that are not associated with violations, and I've learned some solutions. Older 1900s take different mud rings than modern ones, and old mud rings, like that shown in Fig. 2-24, were not tapped. However, boxes of that era contained fixture-support hickeys; I can use one, with the right extension, to support a luminaire's mounting strap. I have a stockpile of assorted extensions, straps, and similar hardware. I show a sample of them in Fig. 2-25.

Closing Reflection

Old work is worth learning, and this book filled a gap in the materials available to electricians. Still, a reminder: this is not a "how-to"

Figure 2-24 Old device rings (mud rings) don't match our 1900s.

Figure 2-25   Adapters help me mix the new and old.

manual. No book could teach us enough to do the job safely. Supervision and experience are essential parts of training. Moreover, any work I perform on old wiring is a compromise, trading less expenditure of money and mess against less safety. That may be blunt, but that is the straight of it. What's the best way to ensure safety? The only solid advice that I can offer without any hedging is this: rewire from the service drop or lateral on, rather than touch anything that another has done or begun—whether yesterday or long ago.

That was said seriously. Wiring a new, I have had the lousy luck to get bad materials and equipment from the factory (well, new from the distributor), and I have failed to catch mistakes made by tired or careless workers, including myself. But at least I was not gambling on as many unknowns—gambling along with the customer. So, "rip it out" is not a stupid choice, just an expensive one.

Whatever I choose to do, ultimately I have to rely on my professional judgment, not just the books that I have read and the other jobs I've worked on. I have to keep my eyes open and my mind alert. This is important, not just for my sake but for the customer and for the next person on the site.

At the end of the day I want to feel reasonably confident. "At the end of the day" has become an idiomatic phrase, equivalent to "when all is said and done," but to me it carries an important warning. If at the end of a work day I feel exhausted, but I keep going because there isn't that much left to do in order to finish, or because I want to keep a commitment, I've learned I can be making a BIG mistake. When I'm wiped out—usually body and mind, but either one—I'll overlook things and make mistakes. I've made as simple a mistake as forgetting to restore a safety disconnect, with the result that product got damaged—so, I had to make good.

So, at the end of the day, I need to have enough capacity left to look back calmly and clearly. Sometimes, this has meant cutting my work day short. When I can't handle everything I want to take on, or I think I'm getting in over my head, I have to pull back. I'm learning that one continuously.

CHAPTER 3

Troubleshooting

Overview

As you read this chapter on troubleshooting, it is particularly important to keep in mind an earlier warning. I have not designed this as a how-to manual. I can't tell you what to do, and doing so is not my intent, for several reasons.

- I am not looking over your shoulder.
- I don't know your background or your standard for risk-taking.
- Frankly, I can't take on the liability that telling you what to do would bring.

I hope that I can give you some idea of what, sensibly or foolishly, I actually do. Taking it from there, you have to rely on your own training and experience, and on local resources.

I use troubleshooting as a framework for illustrating various issues that I face in working on older wiring. Because troubleshooting is so large a part of old work, this is far from the only chapter in which I will discuss troubleshooting.

Safety has to come first. This material is placed right at the front, because it is dead critical and absolutely warrants "overlearning"— which means reading through this even though you may be fully familiar with it and even though the previous chapter focuses on the issue. Next, I discuss the best ways I've found to approach a troubleshooting puzzle. After that, I talk about my approaches in taking outlets apart for examination and testing. After going over how I track down problems, I talk about how I repair some of them. I close with observations on bonding and grounding.

To avoid breaking the flow of the chapter proper, I'll only glance over two oddities: older fluorescent fixtures and old splices.

A Heads-Up on Safety

Here are two important points about safety. In old wiring, even when the workmanship shows that the original installer was a professional, I dare not assume that he followed modern rules. For example, the white or once-white wire is not necessarily the neutral conductor. In old 120-volt wiring, even more than in other wiring, I cannot rely on the return conductor to be an identified, or white conductor. In some old wiring, the return may be energized—even when I have turned off the breaker or unscrewed the fuse.

I rely on both my testers and my senses. I sniff for burned bits and listen for crackles. When working around wires that could arc with significant energy, I need suitable eye and face protection, appropriate clothing, and sufficient distance. What I mean by enough distance, when the energy level is low enough that arc blast and flash are not issues, is room to pull away. I also want a stable place to set myself; so, I'm not wasting energy on perching awkwardly, will not slip, and will not even be likely to hurt myself if I should flinch. Finally, if I'm too tired or distracted, I can't do the work right—or safely. Besides safety considerations, looking and listening and sniffing take alertness and these are important tools for tracking down evasive problems.

Using All Available Evidence

Example   This is illustrated by my customer, Nan, who took a tip on using the nose as a test instrument. I was hired to track down the reason a circuit was tripping One of the first things I did was to sniff at the various outlets and switch boxes on the circuit. A burnt smell led me to discover a ground-faulted switch. The evidence of charring in the smell near the box was confirmed when I removed box and cable from the wall. Two months later, Nan called to say that she needed me again. She had smelled the same "electrical burnt smell" at another switch box, one that, in fact, had checked out okay on the earlier visit. "Could it be hazardous? Should we have you over again?" Yes, and yes again. Smart customer. It seemed that a water leak had reappeared, after a heavy rain, and had traveled to a new part of the basement. (As it happened, the problem disappeared by the time I arrived, leaving no clues. So it goes. If she had chosen, I could have pulled apart enough of her system that I could megger the basement wiring.)

Sizing Up the Problem

General Rules

A customer in an old house complains about problems with a light. At first, this might look quite similar to troubleshooting at a fairly new building.

An important aside
I am much happier dealing with malfunctions that aren't totally dead when I arrive. This makes it easier to identify the circuit and to positively confirm that I have killed power when I shut off the circuit.

The first thing I do is to ask the customer to describe just what he or she has observed. Complaints often are heavy on amateur diagnosis but light on detail. I may need to ask follow-up questions in order to learn what the customer actually heard, saw. or smelled.

I find one useful follow-up question is, "Is anything else not working, or not working right?" Another is, "Did this start happening after any particular event?"

When the neighbor seems to have also lost power, I ask customers to call their utility before even having me out.

Even when the neighbors' systems seem to be doing fine, if everything on one leg or phase is out, this may well point to the service drop or lateral. Other times, multiple legs or phases can be having problems, which can be traced to the customer side of the disconnect.

Example   I performed a remote diagnosis for one old customer who sounded as though she'd lost her neutral. The utility representative had come out and told her the problem was interior, by which I mean at her side of the service point. At her insistence, another came out, actually opened the meter can, and found that because of deterioration of the service cable sheath, water had entered and had corroded the neutral connection.

After getting what information I can from the customer's report, I gather data firsthand. I may attempt to reproduce what the customer reported, and I certainly check for simple causes of failure. If a light stopped working, I try the switch, and I check more than one switch in the case of three-ways. I check for the presence of both a wall switch and a pull chain. If nothing happens when I flick the switch or switches, I check the lamp. Doing this is far less foolish than spending expensive time hunting for something more exotic when, in fact, the customer has burned out a light bulb, or, as I have found occasionally, has bought a batch of bad light bulbs. I also been called in when a customer had left a fluorescent tube in place till its ends are not just dark but black, installed a replacement tube incorrectly, or disturbed the fixture so that a tube shifted out of good contact with the terminal at one end.

With a malfunctioning receptacle, I may first make sure that the (portable) appliance that failed to operate is not defective. I also make sure the receptacle is not switch-controlled—perhaps only in one half. On one occasion, I found the receptacle was fully functional, but the customer had been too weak to push the cord-connector's prongs in far enough. I offered to replace it with a new device. She decided not to have me do this.

I often suggest each of these steps to customers over the phone, if they sound as though they will be comfortable taking them on. Has a fuse blown? I do remind them, though, that if an overcurrent device has operated, it was for a reason—one that is worth identifying.

Other "usual suspects," are as likely to cause problems for people dealing with old as with newer wiring. These include, but certainly aren't limited to, recent worker errors and oversights; interior water damage; and overheating, evidenced by smoke or smell, if at all recent. Even when some of the "usual suspects" are present, I try to avoid drawing a premature conclusion. Many problems in old wiring have had nothing to do with recent work, for example, even though it was performed by an amateur.

Once I have eliminated the obvious, I put on my thinking cap: what clues do I have? I will proceed in different directions depending on whether the light takes a while to come on, flickers when I touch the switch, or sometimes will not come on for days.

If a fluorescent takes a while to come on, perhaps it is a trigger-start that has a dying starter. If it flickers when I touch the switch, this is an easy one—the problem almost certainly is at the switch. (I have learned, though, that I can momentarily make a fixture work that was inadequately grounded just by touching it! Even if I'm not a good ground, I offer a sink for capacitative discharge.)

> While younger readers might never have heard of these, old fluorescent ballasts, even for indoor use, were not thermally protected. I have not run into one of these for some years, but even if they no longer are reaching end-of-life and starting fires, removing one that had gone bad a long time ago could mean exposing myself to nasty pitch that had oozed from the ballast when it overheated.

If a light will not come on for days, sometimes the complaint means that the customer doesn't realize—or remember—that the light is controlled by multipoint switching and that one of the seldom-used switches has been left off most of the time—or is bad. When I've found two three-way switches, I have once or twice found it useful to continue searching—for four ways. The light will not come on until that bad switch is flipped again into its conducting position—usually it has one.

Fatigue Often Is the Villain

Besides any specific clues I come across, there are general rules that can guide my search. What is most likely to have deteriorated?

Deterioration, whether of insulation, conductors, or mechanical components, generally results from fatigue or corrosion.

Fatigue is one result of repeated movement. Heat is a common cause. Thermal expansion and contraction are among the most important means by which heat causes deterioration, but it causes damage in other ways as well. Changes in materials because of heat, however, are not by themselves the most likely source of an old-wiring problem—unless there has been severe scorching. Fatigue caused by thermal or other movement is the most common culprit I find, after careless or ignorant workmanship.

Switches and receptacles are likely suspects, because of repetitive movement over time.

Because fatigue is the result of movement, the first place I look is anywhere something moves, such as a switch. I don't ignore corrosion because of non\movement. Movement can rub away corrosion that otherwise would build up, sometimes freezing mechanical movement and sometimes causing high impedance between conducting parts.

The problem with switches is not just that the devices themselves may die or even that the terminal connections can loosen or otherwise deteriorate. Each time the handle is flicked, the attached wires can suffer a tiny bit of jiggling. I have also discovered terminals that shifted close enough to grounded enclosures, as the yokes shifted under their mounting screws, that the devices arced. Less commonly, I have come upon a similar problem originating from the opposite direction: the box. Either it was squeezed into too small an opening to begin with, or someone inadvertently distorted it when working on the wall. Whichever the cause, its sides were forced in toward the device it enclosed. Finally, while old twist-and-solder terminals stay on, sometimes wirenuts come off. I find this less true with good pretwisting and professional selection. Amusingly, a very early wirenut (a bare conical spring) showed exactly how well you had grabbed the wires. (Tugging at each wire in each splice does the trick for me.)

Smaller movement takes place at receptacles. Worse, receptacles commonly are daisy-chained, and, particularly when using backstabbed connections, here's where I often find downstream continuity interrupted. (I've found this same problem at those backstabbed switches designed with two holes at each end.)

Skillful old-timers would successfully (and legally, at one point!) loop two wires under one screw or, especially, loop around one switch and move on to the next. I find imitators' work not only illegal but unsuccessful, especially over the long term. Looping from one to the next, while not forbidden, is tricky to do well. Then, there are cases where installing—but *not* looping—two wires under one screw is an intended use that conforms to Listing. Figure 3-1 shows a receptacle with only one screw per side. You'd think this would prevent legal

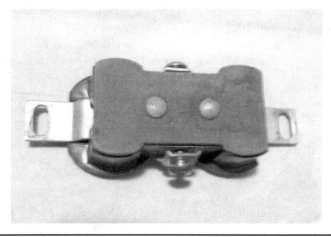

Figure 3-1 Duplex devices can have one screw per side.

daisy-chaining. However, it is a pressure-plate terminal with an opening under each side of each screw.

Lighting Outlets Are Frequently Trouble Spots

Lampholders for incandescent lights are suspect #2. I have more than a half-dozen reasons to suspect a luminaire, and specifically a lampholder, at least when only a single light gives trouble.

- There is gross physical movement every time a lamp is replaced.
- Contact between the terminals and, respectively, the shell and center contact commonly is made through rivets; these loosen or corrode.
- The switched lead usually connects to a vulnerable center contact. This contact either is spring metal, which flattens over time as it loses its resilience, or, in some older lampholders, is a solid, very short, cylinder, which I've sometimes found pitted and offering very high resistance.
- Lampholders and their elements are subjected to long hours of cooking over 60, 100, and more watts of electricity. More precisely, they cook over whatever percentage of those watts the luminaire—often a rather inefficient device—fails to radiate away.
- Overlamping is common. Even if the luminaire is not overlamped presently, it may have been in the past.
- While non-IC recessed luminaires may specify clearance from thermal insulation, this requirement *very* often is overlooked or ignored.

• In old houses, the boxes above ceiling lights often serve as junction boxes, as well as outlets, and are quite crowded. For this reason especially, lighting outlets—both those giving trouble and others upstream of the trouble spot—often warrant investigation.

My general suspicion of luminaires is largely based on troubleshooting experience. Also, in a particularly old house, a gas sconce may have been repurposed for use with electricity. (I talk more about this in Chapter 11, Grandfathering, Dating, and Historic Buildings.) In the early days of house wiring, power was brought to one point in a room—a pull-chain lampholder at the center of the ceiling. Later, receptacles were added by branching out from this often already-crowded box, without consideration of any need for additional volume. The outlet has been crowded for a long time. The wiring in it has expanded and contracted and baked because of incandescent heating for as long.

My Common Order of Approach

Given what I've discussed, I have a general sequence in which I proceed when I explore a "light-out." My first suspect other than a blown fuse or tripped breaker because of recent work or simply overload is, nowadays, a tripped ground-fault circuit interrupter (GFCI). Next comes the silly one—a burned-out lamp. Next is the switch. This includes auxiliary controls, such as motion sensors, which I've seen retrofitted to control some vintage lighting. After this, even if it's inconvenient, I am inclined to explore the lampholder, and even the wiring inside the luminaire's outlet.

This assumes that nothing has flagged my attention. Amateur work, and even indications such as that shown in Fig. 3-2, where the older BX was stapled but the newer was allowed to hang, are possible trouble spots.

Because heat rises, I am more likely to suspect an old ceiling outlet than an old wall sconce, when other factors don't draw me elsewhere. Very new luminaires in old houses also draw my attention, simply because so many replacements are incorrectly installed, or, more important, inappropriately chosen. Not only does the can in Fig. 3-3 lack the required clearance from surrounding asbestos (not being Type IC), but it is fed by BX that illegally stops short of the j-box. Being over 7′ AFF, it might not have required grounding—except for NEC Section 110.3(B)'s requirement, as its instructions almost certainly called for grounding, and no one was around to show me that this wasn't part

Figure 3-2   Good stapling examples rarely influence later work.

of its Listing. Because of this unprofessional work, the splices were strong suspects, and so I considered them worth checking.

I do find that wall sconces often sit on very shallow boxes, too often recessed; and in wood paneling...that can be a bad one. I see this same hazard with receptacles, such as in Fig. 3-4.

The Effects of Deterioration

Simple deterioration because of heating and cooling is less likely to cause a short circuit than to cause an open connection. This includes the possibility of an intermittently open or even an arcing connection. Air is such a fine insulator! When insulation dries, it does not crumble unless it is moved. Even cracked, without movement it still keeps conductor apart. If it does crack and crumble, the conductors that were protected by the insulation have to be moved against a grounded surface or each other—or moved very close—before they will short.

Figure 3-3
Dangerous thermal insulation may hold bad wiring.

Figure 3-4 Siding in enclosures is a common hazard.

That is a lot of movement to result from heat expansion, or even from screwing and unscrewing lamps from a lampholder so long as the lampholder itself does not twist. Often, of course, lampholders do twist, a problem that I will get to shortly.

Natural-building movement, as well as vibration from activity on the floor above, can and does nudge wiring. I've seen early failure of wiring chafed by the machinery it fed. However, neither seems nearly as likely a cause of wires touching as is someone having opened the enclosure and handled or nudged them. This makes me especially leery of exceeding the limits of old wiring. I know that my troubleshooting could, in itself, tip a shaky system over the edge, and bring on a short (or an open) that hadn't existed before.

When this type of arcing does occur, Earl Roberts reported, tests showed that GFCIs did not prevent overheated fixtures, recessed in insulated ceilings, from causing fires. Damaged wiring can cause a fire long before there is sufficient arcing to ground to operate a GFCI. (This discussion took place pre-2006, but I don't trust the situation to have changed with the updated GFCI standard.) Arcing fault circuit interrupters (AFCIs) do rely on GFCI/ground fault protection of equipment (GFPE) circuitry to interrupt power to faults caused by arcs that have not activated the AFCI-sensing elements. Unfortunately, an AFCI might not trip in response to a high resistance, low-amperage arc. I was warned about this at the time that AFCIs protected from only parallel faults, so, it may have changed in combination units. High-energy arcs that burn free quickly are highly unlikely to start wood burning in residential contexts. Other settings, where the energy levels and available fault currents are much higher, can be a great deal more dangerous.

> **Note**
>
> I'll repeat this later, because I found it an eye-opener: Ideal's top testing expert, Brian Blanchette, reported that the AFCIs he exercised to failure had only 75 trip cycles in them. After this . . . no tripping in response either to test or to actual fault.

I commonly trace shorting to lampholders where something pushed the wires against one another and severe aging or, commonly, further twisting or pushing wore away the insulation. If the lampholder is not solidly secured to the fixture, the lampholder may rotate each time a lamp is replaced, twisting the wires behind it (see Fig. 3-5). The image shows stressed-to-failure THW wire, whose insulation is tougher than the old rubber-insulated conductors—but still was destroyed by lampholder twisting.

Bezel rings often loosen with time. However, the type of lampholder I find most prone to this problem is canopy-mounted. Two

Figure 3-5 Twisting damages even stranded THW.

porcelain components screw together and are held from rotating by engaging with a bump in the thin sheet metal on which they are mounted. Sometimes this is not much of a bump, and thus not much of a restriction.

Energized Testing Procedures

I've seen this problem traced down in nonenergized wiring by a physicist customer using his micro-ohmmeter. However, I'm more familiar with energized testing. Obviously, this approach can be far more dangerous. It probably would have killed me by now if I simply ignored the drill on safe voltage testing.

This said, I like to start at a known source of power and work my way downstream. In low-voltage work (the bulk of my troubleshooting experience), I usually take an extension cord with me, offering proven energized and neutral conductors.

Consider the nonfunctioning light fixture I discussed near the beginning of this chapter, under "General Rules." Suppose I test the lampholder for the energized (presumably switched) conductor and the neutral conductor and find both present. If the lamps show internal continuity, I have determined that there is a contact problem that allows my voltmeter to show a complete circuit, but does not pass power through to lamps. Often, it is caused by corrosion or loosening. On occasion, the glass bulbs on replacement lamps—what should they have been: A19? A21? PAR20, PAR30, or BPAR40?—have had too-wide shoulders, preventing them from screwing in far enough to reach even a healthy, energized center contact.

Often, I've eliminated these possibilities and rechecked and the light worked fine. There has been an intermittent malfunction in the switch, even though it tested out okay. I also have found intermittents further upstream, even at and in fuseholders. The nastiest one I uncovered recently was caused by a new-style BX connector leading out from the distribution panel; its armor stop penetrated conductor insulation. Sometimes, there has been a problem that was temporarily reversed by the twisting associated with unscrewing the lamp. I usually have to start opening enclosures, starting with the one at the light.

Playing It Safe(r) as I Pull Things Apart

I next remove the fixture or device or at least pull it away from the outlet box. Yes, I have performed this work energized, even when I knew how to kill power; but when there's no need, what do I have to prove?

Proceeding to Explore

When I don't know what circuit is involved because an outlet is intermittently dead, if I remove the cover plate and find power at the switch, I am ahead of the game. If power comes up to the light through the switch, I'm somewhat inclined to look there first for a source of intermittency, for several reasons. For example, in terms of ergonomics and efficiency it makes more sense to work at arm height rather than from a ladder, other factors being equal. Also, the wiring at a switch is likely to be less fragile than at an old light. In addition, it may be closer to the source of power.

However, this may go either way. Other reasons may convince me to start at the lighting outlet itself:

- Sometimes I have needed to get up on the ladder anyway, to check the lamp, and proceeding from there has looked straightforward.

- Sometimes I've been earnestly trying to avoid disturbing the expensive wallpaper abutting the switch plates.

- Sometimes the light fixture had exposed wiring; a chandelier is an example. The exposed insulation was visibly ratty.

It is a reasonably safe bet that going to the loadcenter and turning off the power feeding the switch will interrupt the power trying to get through to the light. However, I have been surprised when I tested. The most common reason is that the obvious switch doesn't actually control this light. I also have found multiple circuits represented at a switch, although much less frequently than at a light.

Even worse than the presence of additional circuits is the following hazardous scenario. I check for voltage between the center contact and the lampholder shell and find nothing. I confirm that the outlet is dead by checking for voltage both between the center contact and the outlet box and between the shell and the outlet box. Nothing. Just to be safe, I test "everything against everything" including my extension cord. This is not overkill.

I have found the return path interrupted and the shell used as the energized or switched-energized contact—and the outlet box not well grounded. Or the center contact is the energized contact, the outlet box is a bad ground, and the shell is backfed from another energized circuit. A bad ground is not the same as no ground. A little pressure on the box has caused it to make better contact with a BX connector; other times, starting to loosen a screw cut it free from the paint insulating it or scrubbed corrosion away, putting more conductive surfaces in contact. Then, suddenly, the canopy or box is well-enough grounded to shock me when I touch it and something energized. Here a noncontact voltage sniffer *might* help. However, I trust sniffers, like basic LED receptacle testers (bugeyes), when they warn far more than when they reassure.

Backfeeds that cause the shell to be energized, even when the fixture is wired correctly, can have various causes. One involves a neutral conductor that is broken upstream—perhaps, in some very old buildings, by a blown fuse. Neutrals were fused, along with energized conductors, for some time. For example, Section 807 of the 1925 NEC authorized inspectors to require both. The power is fed into the conductor downstream of the break from a load on a different outlet on this circuit or from another circuit properly or improperly sharing its neutral conductor. Consider how backfeeding can result in double or in zero voltage between my two conductors—because they are both energized. The lighting outlet on circuit 25 could operate on up to twice the normal voltage, because with the break in the neutral conductor, it only can complete its circuit by a path through the receptacle and circuit 28, dividing phase to phase voltage, or energized leg to energized leg voltage, with whatever load is plugged into the receptacle. If this were not a legitimate multiwire circuit, if circuits 25 and 28 come off the same leg or phase, it could not operate at all without a neutral conductor. This could happen because someone just grabbed any empty space to feed a circuit, or because this is one of those odd old panelboards where it is important that workers read the schematic. Some of these will show that the busbar feeding circuits 25 and 28 is the same; in others, circuits 25 and 28 are not where one would expect them—the numbering scheme is not the typical one. I show pictures of such a panel in Chapter 6.

Another cause comes from a switching system that totally ignores the "center contact–energized contact" convention. This is discussed in Chapter 5, which addresses outdated switching designs.

More and more, I will kill all power to the building. First, I warn the customer of the need to reset most equipment with clocks and timers, and also check that:

- Computers are turned off, unless they have uninterruptible power supplies with ample batteries.
- Important processes—from fine machining to laundry drudgery—are not abruptly halted at awkward points so as to create danger or unnecessary expense.
- Refrigerated perishables are not put at risk.
- No one is endangered by being caught in the dark. One customer who lives in an old house has such bad vision that his maneuvering is very poor. This can put him in danger, especially in emergencies. However, he is not prepared to have me install a backup power system to ensure adequate light, or even improved lighting for daily use.

Unfortunately, even operating the main service disconnect is no guarantee of safety. In some instances, even pulling the meter is not enough. With old wiring, I can be even less sure than usual that when I think

I have killed a circuit, it is dead, or that when I think I have shut down everything in a building, all power is off. I always have to test. I always assume that I am dealing with energized wires until I prove otherwise.

Despite the fact that there is no absolute guarantee that I can kill all power, I need to locate the circuit on which I am working, both for safety and to ensure that the problem is not one of a tripped overcurrent device. I use certain clues to shortcut my search. If I'm troubleshooting a light, I try to find out what controls nearby lights and receptacles. This gives me a place to start. In a commercial setting that has one light out, if there are rows of lights that are independently fed—but that are inadequately marked—I look for a breaker that feeds one light fewer than the others seem to be handling. Sometimes, but not always, that is the circuit that controls the problem fixture.

Whichever wire is no good, the other or others may be continuous. With 120 volt, I use my extension cord to determine this, after going through the "test my tester" safe-testing preliminaries. This way, I can find out how to disconnect the still-conducting wires, whether neutral or energized, at the source. By virtue of the cable or raceway they leave in, I often know which wire should be completing the circuit. A suitable circuit tracer could do the trick (although most of these are not designed to handle three-phase systems, according to testing expert Brian Blanchette). I *still* would want to double-check. A warranty wouldn't compensate me for electrocution.

When the source of power does not come through the switch, which merely contains a switch leg, or, likewise, when there is no wall switch, this troubleshooting has to happen at the light. If I am able to confirm the identity of the cable or raceway, I can pull the fuse or flip the circuit breaker feeding the energized wire or wires. If I've disconnected the return and it turns out there's a mess of circuits involved, I hope I get all of them. I certainly have been fooled. The existence of the tangle could mean someone indiscriminately spliced together the neutral conductors of various circuits at some junction box. Before I can consider the system halfway safe, they need to be untangled. The resulting neutral loop can result in very high electromotive force (EMF)—but only while at least one of these is feeding loads—as well as creating the confusing need to kill multiple, otherwise unrelated circuits to eliminate net current. Sometimes, this was simply the result of "white goes to white" thinking. Other times, this was a bootleg workaround. I may find a broken, or even a burned-up, return. I've also found a white wire interrupted by an unnoticed switch, which might have been unnoticed because the troubleshooter didn't realize that a different switch, that also controlled the light, wasn't operating alone.

Before I disconnect them, I have to check out all the energized conductors associated with the wires I will disconnect. This can be tedious and, sometimes, nasty. The main reason for doing so will be to figure out which energized conductor, and, hence, which overcurrent device, might serve the nonfunctioning outlet. The other reason is that I do not want to risk interrupting the return current from a load, which would mean that I would become part of the circuit. Killing power and using a bootleg is a quick alternative that I won't consider; it seems to be what led to some of these messes.

A ground return is among bootleg options the installer may have used. If this wasn't a quick fix for a problem that I can, in fact, track down and solve, I need to rerun the wiring to make it reasonably safe.

Nonresidential note

IAEI colleagues tell me that they have found bootleg grounds common in higher-voltage lighting fed by power panelboards. Maintenance men, in their experience, are responsible. As these nonelectricians attempt to troubleshoot the nonfunctional lights, a wire makes contact with the panelboard frame, and lighting is restored. The quick-and-dirty solution strikes them as obvious. They make a permanent connection, the return voltage makes its way back along the conduit, and eventually there's a fire.

Fragility   I warn my customers about the risk of damaging building structure. I minimize unnecessary damage by using some, but only some of the same tricks I apply when working on newer buildings. I have found that banging equipment to free it, for example, poses a greater risk to older, more fragile wall and ceiling surfaces. Like the structure, older fixtures often—but not always—have proved more fragile than new ones.

Here are just a few of the fragile bits I have discovered:

- The separator between the actual socket and a metal lampholder, which basically is paper or light cardboard, has dried out and crumbled.

- A cracked, or at least crazed, glass fixture part such as a shade is ready to break apart when handled.

- Similarly, a wooden fixture part has dried out to the point where it is ready to disintegrate.

- And, of course, rubber or cloth parts turn to dust.

Taking Apart the Fixture   When I am not absolutely certain power is off, I'm extremely careful in this part of the process. This is when I have turned a bad ground into one with sufficiently low resistance to draw an arc or shock me. That is a terrible way to find the circuit. It's also happened that I twisted or pulled a conductor so a fatigued wire broke, or deteriorated insulation fell away. Figure 3-6 shows one

FIGURE **3-6** Pliers keep my fingers clear of short wires.

of those boxes whose conductors I dealt with very cautiously, where while I pulled away the wirenut only when I had reason to believe it was deenergized, I still didn't risk putting my fingers in there. This is only in part because of the broken-off wire in back (see Fig. 3-7). I'm especially wary in old buildings with metal ceilings, despite their being coated with paint.

I incurred a multi thousand-dollar chargeback (voluntary, but it's part of my way of doing business) when I left a circuit off long enough that a customer's product deteriorated. He didn't notice, I didn't notice. Ouch.

Proceeding Inside an Old Outlet Box

I consider these issues as I unscrew the screws holding the fixture to the ceiling or to the outlet box, or unscrew the bezel ring to get the canopy down. Sometimes, the source of the problem—or one source, unfortunately, with another lurking—is obvious. A bad splice loosens further or comes off. Generally it is worth my while to take a quick look for additional problems.

Even in the easy case where the fixture comes away, leaving the wire ends in the box, questions remain. How did the splice manage

Figure 3-7 Hidden wires may be energized.

to conduct power to and from lamps as long as it did? Perhaps the broken ends maintained just enough contact.

In this easiest case, I found the wires broken off right at the lampholder. Other times the problem has not been so evident that it yielded to a quick visual inspection.

I have not yet heard of a wire that was in use simply breaking inside a cable, in the middle of a run, after being buried somewhere within the wall. I have learned, to my surprise, that staples pounded into cable, however amateurishly hard, are highly unlikely to be responsible for, at least, fires. The exception is the staple bent over so the sheath edge, rather than the flat side of the cable, was rammed into by the staple, as in Fig. 3-8, which includes barbed-wire staples misapplied to NM. Inspector and instructor C. Larry Griffith found evidence that a staple over on its edge started a fire. It was an old, nonrounded staple, making it easier to cause damage. Unlike the one in Fig. 3-9, it hadn't been driven even more-or-less flat and the cable had been hauled against its edge, presumably by someone trying to draw it tight to remove slack in order to gain a few inches. I would hazard that a modern staple that's bent over, such as the one shown in Fig. 3-10, badly bent,

FIGURE **3-8** Barbed wire staples are misapplied to wiring.

is a greater departure from its Listing, offers poorer support, and may pose more danger in the event that safety hinges on its compliance with the Code requirements for supporting and securing cables. A staple gun staple may not pose much danger of squashing the cable when attached to the edge of NM, as shown in Fig. 3-11, rather than

FIGURE **3-9** A staple that's mashed, but flat, usually is still okay.

FIGURE 3-10 Old staples mashed flat beat new ones bent over.

the flat, and it may not pose much risk of penetrating the sheath and insulations. On the other hand, it certainly has less chance to penetrate as deep as possible into the joist for support.

A clamp, too—such as in the box shown in Fig. 3-12—may be designed to bear against a flat surface. As shown in Fig. 3-13, as I see too often even in permit-covered work, this particular clamp was destroyed. The discussion of damage from staples, which are at least

FIGURE 3-11 Staple gun staples have to penetrate fully.

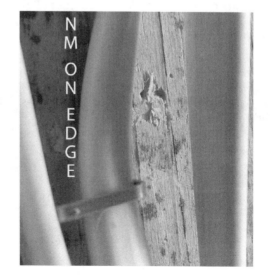

FIGURE 3-12
Self-holding boxes
seem perfect to
laymen.

Listed for cable support, says nothing about bent-over nails, or about plumbing hangers, such as was used in Fig. 3-14.

Certainly there are defective cables including some that are sold with voids inside the conductors' insulation—and I mean actual gaps in the metal of the conductor. However, all that have come to my attention failed to function from the moment they were installed. (I caught one in 2009 in the process of installing new NM; luckily, the void was close enough to the sheath end that I recognized it without actually having had to strip back.) There are cases where a cable gets pierced by a nail or screw, which cuts and sometimes opens or shorts a conductor, but those are rare exceptions among the problems I encounter. AFCIs were developed in small part to deal with these rare problems. Because wires almost always fail at or very near connections, in modern legal wiring they die inside enclosures, or at least

FIGURE 3-13 Many
don't understand
plastic clamps.

Figure 3-14 I find plumbing straps used on NM.

at accessible locations. Even in older and less legal installations, I still find this to be the case, not because cable is run correctly but because it is so easy to make bad splices and terminations.

Sometimes, wires are nicked in the original process of stripping off the insulation. Sometimes, they are shoved back and forth inside the enclosure. Sometimes, they are twisted and untwisted as splices are made and remade. At outlets, as opposed to junction boxes, they almost always are subjected to the heat of being close to where electricity flows through assorted impedances. These include not only the resistance of adjacent splices but the impedances of hardwired loads such as lampholders plus lamp filaments located below or near the outlets, and even receptacle terminals, prongs, and plugs. Switch boxes are intermediate between junction boxes and outlets in the heat stress they put on connections.

Now that I have pulled things apart, before I assume that there is a hidden problem it is worth my while to retest. Practically speaking, this means restoring connections at the loadcenter if I have them disconnected or turned off. If possible, I will disconnect the power feed again before digging into the wires as I proceed further. Even so, I prefer to use an insulated tool if I have any suspicion that a wire may be energized.

Safety perspective
For my own purposes, I do treat the clean rubber or plastic comfort grip of an ordinary electrician's hand tool as though it offers some kind of insulation, even though it is not Listed or rated as serving the purpose. It's as though I were using dry cotton, or wood, which similarly has not been investigated for the purpose. If this sounds perfectly good enough to use when actually working live, I suggest you flip ahead to Chapter 6, and read the example of Jim Wooten's discovery, a few paragraphs after the heading, "Mystery Voltage."

Troubleshooting Old Splices
To test for voltage (or for continuity), I may need to get into the splices. Very little is likely to go bad with old, soldered-and-taped splices. The conductor insulation may well be brittle, as in the case of the insulation in Fig. 3-15 (at least 60 years old), but that bad insulation, not the connection, will be the strong reason to consider rewiring. True, splices have gone bad inside the tape, where wires have started making bad contact inside the splice. However, I've found this to be exceedingly rare.

Broken Wires   On other rare occasions, fatigue has actually broken one of the conductors outside the splice itself, but just *inside* the tape! Wires subjected to a lot of fatiguing movement have broken off inside their original insulation at other places. Just like thoroughly fractured or dislocated bones in a person's arm, these may manifest their presence by making sharp angles when moved. It is easy to confirm the break; old insulation is not strong enough to keep me from pulling it apart

Figure 3-15   Old rubber can crumble away.

unless there is a wire inside to give it some backbone. Similarly, wires have broken off just at the screw terminal of a lampholder, switch, receptacle, or fuseholder. I touch these and they move away.

One more way I have checked for breaks, as well as for faulty splices, when certain that it is safe to do energized testing of rubber-insulated wires, is poking through the old insulation with the sharp tip of my voltmeter probe. With power off, this means my ohmmeter or continuity tester. A wrap or two of tape has made up for any injury to the insulation. I emphasize that this is for soldered-and-taped splices. With normal, untaped wirenuts, my probe can reach in and touch a wire or spring.

Example   An 8B contained a pair of somewhat-black wires (I decided their splashes of white paint almost certainly were unintentional), spliced, and tucked away in back; a pair of somewhat-white wires, spliced to the white lead of a nonworking lampholder; and another somewhat-black wire, which was attached to a black lead going to that lamp. The insulation was healthy-appearing rag wiring, and the wirenuts were old-style rectangular Bakelite ones, thoroughly covering the wire. I thought I'd killed power to the box, but wasn't very sure.

What I wanted to do next was to determine whether power and neutral had been reaching the lampholder. It seemed safer to remove the wirenuts than to try to pierce the insulation. Which of the two was more dangerous to unscrew?

Unscrewing the wirenut attaching the lampholder to the single black wire, I kept the wire carefully under control. I tested it against a good ground and found no voltage. As I was unscrewing the wirenut attaching the pair of probable neutral conductors, the splice broke. One of the wires sprang away, and—they were, in fact, neutrals—it turned out to be carrying the return current from some load. I had just interrupted the path this current had been traveling. In modern wiring, this problem is more likely to appear as the result of removing wirenuts holding together neutral conductors that had been spliced without pretwisting. On the other hand, it seems more common, in modern wiring, for my probe to reach something it can test without my having to loosen the wirenut. However, I trust positive results of such testing far more than negative. I don't even fully trust a no-voltage reading when I probe a single wire that is part of an uncertain-looking splice.

Example: No Power at the Outlet   Sometimes the source of power does not come through the switch, whose enclosure merely contains the feed and the switch leg. In other words, the neutral conductor was not brought through the switch box. I find no power at the switch. I find no power in the lamp socket, nor do I find power at the terminals at the back of the lampholder. (I need to check at the lamp, because it is always possible that the installer, or somebody messing with the system subsequent to its original installation, arranged that the switch broke the neutral rather than the energized conductor.)

This example is relatively simple only when I apply the label "simple" in contrast to the far-worse messes this work brings my way. If you want, you're welcome to try to diagram the following; but I present the example as a typical nightmare rather than as a lesson in how to troubleshoot.

The light I am examining is the only one controlled by its single-pole, single-throw switch. There are only three splices in the box at the lighting outlet, to which three or more cables are connected. Because of the general appearance, I trust that there are no insulated grounding conductors.

Here, is what I find as I track the wires. One splice contains two conductors: one of those heads out of the box in a first cable and the other goes to the lampholder center contact's terminal. I may assume—but I always need to check such assumptions—that the first splice constitutes power returning from the switch, the switched energized lead. (If I can be certain which switch is involved, I know where I can confirm this, but only if I have the right switch and the conductor has no open.)

The second splice contains three (or more) conductors; one goes to the lampholder shell's terminal and the others head out of the box in the other cables. Ideally, this second splice contains the neutral conductors. (This is another assumption that I need to confirm.)

The third splice contains a number of conductors, none of which goes to the lampholder, but one of which seems to leave the box in the very first cable, the cable I suspect of going to the switch. This certainly should be the splice containing the energized conductors coming from the source of power and continuing downstream. Even if it turns out not to be the right splice, it certainly is worth my while to test it for voltage, if energized testing seems safe. Given that I found no power at the lampholder, and given that the other two splices connect to the lampholder, where else could I look within this outlet? Since, unless I have encountered a Carter-system scenario (see Chapter 5), one splice seems to contain the switched conductor, the other should be the return, and the third ought to be unswitched power. Hence, the third is worth checking as possibly constituting the furthest upstream point at the outlet itself that I can check for power.

It's not usual for me to get essentially a zero voltage reading at this third splice. This means either that there is no power at the box or else that it was not wired the way I'd hoped. I need to check somewhere else. With solder-and-taped splices, probing each conductor going into the other splices can be at least as useful as probing the splices themselves. Unless I am pretty certain that the circuit is dead, or am pretty clear as to which outlet power is coming from—which should be the closest point upstream—and can check to see if that one is dead, my best bet is to assume that power is getting to the box, but not making it through some single conductor or connection. This working hypothesis means that the first thing to check for is a conductor that does have power. Then I look for the break.

Evaluating Bad Wires

What do I do when I find broken conductors or bad insulation? The problem is not all that different from the one I face when I am simply replacing an old fixture. I have to undo and redo the old splice. Often everything's too short. At best, conductors may be barely long enough, but need to be shortened because the wire inside is damaged or, alternately, the wire's surrounding insulation is gone.

I'll explain what I take these terms to mean. A damaged conductor is one that is nicked or in some other way reduced in diameter. I take insulation to be essentially "gone" when it fails the following test, rather than by relying on a megger's snapshot. I perform a rigorous test of insulation condition by folding a conductor double and straightening it. If this does not exceed its limit, I figure it has a few years left. By reaching its limit I mean giving one of these two indications of fatigue death: cracks in the insulation that are evident to the eye, or a conductor that breaks, evident because it either visibly separates or else it makes a sharp bend, like a broken bone under the skin. I recognize that any such testing, being a stressor, will bring marginal equipment much closer to failure.

Healing Old Wires

Sometimes, the wires fail my test. I could throw up my hands and say the whole place needs rewiring, or I can offer choices that may be less ideal than rewiring, but in some cases are more realistic, given the customer's limitations. The limitations may involve ability to budget the cost, or even to tolerate the upheaval associated with rewiring. (See Chapter 9 for the options I have offered.)

Some electricians talk of stripping the insulation off new wire from the reel and slipping it onto old wires. I haven't found this distinctly non-Listed use very satisfactory. For one thing, I still need to tape the transition between the new slipped-on plastic insulation and the remaining old cloth-and-rubber. There are relatively exotic solutions, the most familiar of which may be voltage-rated heat-shrink tubing, applied by someone experienced. I've used it, but don't employ it much. This is simply because I often can make taping work okay, and taping does not involve standing on a ladder with a torch or heat gun. As with any repair, I need to extend the end of heat-shrink over good insulation in order to avoid a gap appearing in insulation as the wire shifts.

FIGURE **3-16**   Tape is not always enough.

Tape is a fine answer only when I can reach in fully and apply it correctly, double-wrapping. Taping old wires can be tricky. The tape can't pull old insulation off the wire and both ends of my tape must adhere to material that retains integrity. Usually this means sticking to the conductor itself, at one end, and to relatively resilient insulation, at the other. If the insulation is dried out too far (see Fig. 3-16), the situation calls for rewiring. When I do tape, I normally color-code the conductors.

Not infrequently, I will find a rag-insulated conductor whose cloth protection is fairly shot, all loose and in the way, but whose rubber insulation itself still has some life. This conductor also gets some additional protection, because the rubber was not intended to do its job without the mechanical protection provided by its cloth covering. This design functioned much like the old splices: wires twisted and soldered, then well-wrapped with rubber tape for electrical insulation, and covered by a layer of friction tape for protection.

Especially when I am dealing with pipe or tubing, as opposed to cable, the fix that I recommend to my customers as the very best is pulling fresh conductors into the box, despite the fact that we may not be in a position to rewire the whole occupancy or building.

Example   I have had more than one customer with just that kind of rotten-all-the-way wiring, but who chose to have me tape rather than replace conductors, because he did not have a lot of cash to spend right then. His crowded ceiling boxes had wires heading out here, there, and everywhere. If I had started replacing conductors, I would have been going at it for days—just to clean up one outlet. Taping the cracked insulation at the one box where he was having trouble may have put a temporary lid on his problem, without disturbing the wires at four or five other boxes. Here are some of the other factors that went into my choice:

- Our communication had stood up to testing, and was quite clear.
- The customer knew that this was not a permanent, long-term fix, and was willing to take the risk.
- I documented my concerns and advice.
- The building benefited from fireproof construction.
- If I had out-and-out refused to take any halfway measure, the customer would have employed a handyperson to try to get things working. While the threat of unethical or incompetent competition has never determined how I work, it is unquestionably true that I do not work in a vacuum.
- Finally, the most important factor of all is that I was able to tape the wires in such a way that I felt confident that they were unlikely to short out as a result, and power could be restored safely.

All this said, I recognize that another reasonable, competent electrician might well have insisted on all or nothing.

Unfortunately, when I am dealing with cable rather than conduit, rewiring is not as easy a matter as using the old wires as drag lines to pull in new wires from my reel. (Several customers have in fact asked me, in all innocence, whether I could pull new wires into their old BX.) As mentioned later, I usually can cut back on the cable sheath or armor if I tussle with it and haul more into the box—or if I replace the box with a bigger one, because otherwise the cables are not quite long enough for me to be able to cut back sheaths to reach good insulation. Often much healthier conductor insulation appears just a couple of inches outside the enclosure, where it has not been subjected to the same heat. In addition, if a conductor has been clipped—this usually has been the grounding wire—I can retrieve enough to pigtail it (see Fig. 3-17, which had to be an extreme close-up to show what was left of the grounding wire).

Receptacle Troubleshooting and Replacement

So far, I've discussed troubleshooting primarily from the perspective of problems with old lights. Regarding the obvious, preliminary tests; it is easy to extrapolate from old lights to old receptacles. Failure

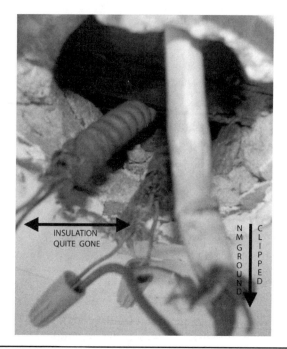

INSULATION
QUITE GONE

N M G R O U N D

C L I P P E D

FIGURE 3-17   I grabbed a quarter-inch to strip sheath back.

often results from incremental deterioration or loosening of points of contact.

When I replace a receptacle in an old wiring system, I watch for a couple of things that are relatively rare with newer systems. One is a lack of color-coding or other polarity identification. I use my voltmeter and, if I feel particularly unrushed, even when I haven't changed anything, I add a little white tape to color-code the neutral conductors, turning them into properly identified conductors. Another common aspect of old systems is lack of grounding. I like to install a positive ground even at a switch or light. I warn customers that people relamping ungrounded fixtures who come in contact with some grounded surface are at increased risk of shock if something goes wrong; they should turn off the circuit. Most ignore this comment, I'm pretty sure, but I've said my piece, giving the advice the amount of weight I feel it deserves.

Bonding Equipment on Which I Have Worked

Before I discuss grounding in further detail, I must consider bonding. I need to take what measures are available to reduce local potential differences. Bonding is not very difficult. A device used to be considered adequately grounded and bonded when it was held to a grounded metal enclosure by a 6-by-32 metal screw. Nowadays, more is required, and I try to bring equipment up to today's standard whenever feasible. While I inspect just to *Code*, and certainly will accept that much old equipment is grandfathered, in my work I prefer to run a bonding wire when a device is not self-grounding. This said, here again I don't ignore the realities of time and money when an alternative, such as solid metal-to-metal contact, is available.

Ceiling outlets in boxes that lack cover screws normally contain fixture-mounting hickeys. Modern luminaires use mounting straps, all of which have center holes that will accommodate hickeys, so the strap can be secured with the locknut that's appropriate for the hickey. Many of these straps have grounding/bonding screws.

With a receptacle, I believe the best solution is to bond to ground wires that are spliced together and secured to the enclosure if it is metal. If there are no ground wires, but the enclosure is grounded, the next best solution is bonding to the enclosure using a 10-32 or larger screw in a tapped hole, unlike the creative installer whose work you see in Fig. 3-18. If there is no prepared grounding screw hole, I can drill and tap one. If there is an unused and unneeded cable clamp, I remove this and use its screw to bond the box, a process I show in Figs. 3-19, 3-20, and 3-21. It is rare for the underside of the screw's head to be rounded, but where that is the case, I need to use a flat metal washer between the screw head and wire. I've never had reuse of former clamp screws rejected. I've also discussed this approach with standards experts.

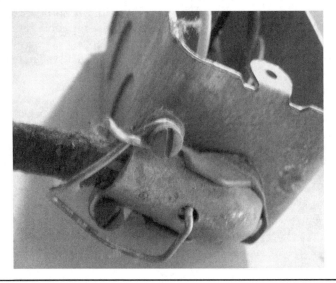

Figure **3-18** An ignorant attempt at bonding.

FIGURE 3-19   First I find an unused clamp.

Even grounding clips can work okay, where suitable for the box. However, it is easy to install them in such a way that they do not make solid contact, or that they are forced off the box as I tuck devices and wires back into the enclosure so that I can install the cover plate. This makes them pretty much my last choice.

Grounding Problems

I find three common reasons for an enclosure to test as ungrounded. One is a poor reading because some barrier is interfering with the tester's contact. Perhaps my tester could not find ground, even though it is present, because of dirt, paint, or corrosion. The solution is to scrape away the interference and bond to clean, shiny metal that now indicates a good ground. If necessary, spray paint will normally restore corrosion protection such as enamel after I make my connection.

The second reason that I run into for a box to test as ungrounded is loose contact between the enclosure and the metal system feeding

FIGURE 3-20   Out with the clamp and back with the screw.

FIGURE 3-21 It's now become the bonding screw.

it. For this reason, even if, on testing, an enclosure appears not to be grounded, I like to test a metal conduit or cable armor for ground. Again, I need to scrape through any paint or corrosion. Armor held by a loose clamp or connector has itself proved to be grounded without having grounded the enclosure. Similarly, conduit has made poor contact. All of these situations have, one time or another, started me out thinking I had a bigger problem than I did.

Then there are unprofessional installations pure and simple. The contact is poor because the installer was unconcerned, ignorant, or inept. In some of these illegal messes, the cable or conduit merely was stuck in through a KO. I add a connector, even where this means chopping open a wall. Sometimes, armored cable is sort of secured by an NM/loom clamp, in violation of its Listing. In other old jobs, someone ignorant tried to secure a locknut against the side of a round box. I have encountered this problem with receptacles less often than with lights, but I have run into it, especially with surface-mounted equipment.

There is no mystery to how I handle these problems. I've tightened that loose locknut or setscrew; added a second locknut, installed the correct connector, in an appropriate opening; made sure the armor is squarely under that armored cable clamp, cinched down carefully. Sometimes I do have to replace a box, of course—especially a plastic nail-on switch box retrofitted to BX.

I keep a few two-hole BX clamps in my truck, such as the one for octagon boxes shown in Fig. 3-22, for replacing those newer ones that will not quite do the job. One reason is that the old clamps may accommodate larger cables than will fit into setscrew connectors, or even the latest AC–MC connectors, clamp, push-in, or whatever the configuration. Another is that a setscrew can push armor right through frail insulation. I tend to be best off when I can dig up one that

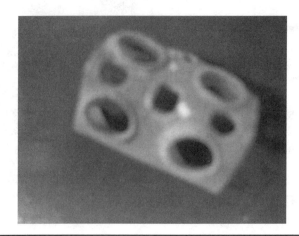

Figure 3-22   This clamp accepts cable from two angles.

really matches the cable and box. In places where I find work whose assorted errors at one location bespeak thorough ignorance, I know from this—and I warn the customer—that the run may be dangerous in other ways. On occasion I have been authorized to check all points of termination along the run. If it is just too bad, I might not be willing to reenergize it "on my watch," because I don't want to shoulder the responsibility for the future behavior of the hazardous mess.

Sometimes, after fixing any number of the problems I just mentioned, I've still found no ground. If everything seemed secure at this end, and the wiring method is one that should provide a ground, I have to hunt to find the other end of the cable or raceway. Somewhere along the circuit, grounding has been interrupted. If necessary, I start at the panelboard and work forward. Sometimes I will find poor contact that developed, as the result of, say, long-term exposure to environmental factors or user overloading. It is probably more common for me to find grounding disrupted by thoroughly ignorant amateur work somewhere along the way. If doing so is at all possible, I correct this poor connection. Some of the situations I find call for rewiring. These include damaged cable or wire, and broken conduit. Figure 3-23 shows work that might have been installed by a contractor, but it fits into this category. Figure 3-24 shows a prime example of how well it can pay to start at the source of power. The intermittent ground resulting from this combination of error could otherwise have resulted in a very frustrating hunt. The sort of dangerous work I found there and in Fig. 3-25 shows the partial and imperfect understanding that I associate with permits pulled, fraudulently, under "rented" licenses.

FIGURE 3-23 Bad locknut use interrupts grounds.

FIGURE 3-24 The more locknuts and connectors misused, the smaller the chance of continuity.

Figure 3-25 Bonding, splicing, and CB misuse often appear together.

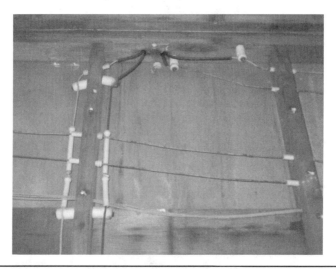

Figure 3-26 Even K&T that was installed well used to lack a ground.

FIGURE **3-27** Knobs use four elements.

What about those cases where I just cannot find an interrupted ground, or where the wiring method itself does not allow for grounding? NM used to lack a ground. Until very recently, knob and tube wiring (K&T) still was a *Code*-recognized method and, traditionally, it was two-wire. The circuit shown in Fig. 3-26, with very carefully done, but ungrounded, K&T is a classy example. It uses ceramic knobs as mounting hold-offs (Fig. 3-27), with leather bushings (Fig. 3-28), under the heads of the mounting nails to protect the ceramic. Nowadays, plastic bushings are used for this purpose. It served surface-mount

FIGURE **3-28**
Bushings cushion
nails in knobs.

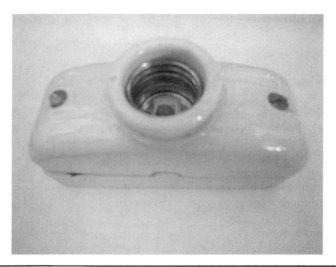

Figure 3-29 Porcelain insulated this self-contained device.

devices (Fig. 3-29, outside view and Fig. 3-30, inside view); nonmetallic boxes, such as in the 1940s, the ceramic switch box in Fig. 3-31, the utility box in Fig. 3-32, or metal boxes fed through protective bushings (Figs. 3-33, 3-34, exploded view, in place, Fig. 3-35, outside view, Fig. 3-36, inside view. Tubes (Fig. 3-37), serve as protection when it runs through structural members, or over other cables. The basic concept

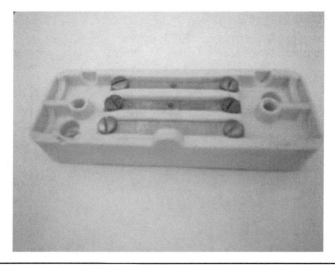

Figure 3-30 The inside view of a self-contained device is logical.

FIGURE 3-31   A porcelain switch box didn't short.

goes back to the early days and is shown in the 1917 motor-wiring diagram of Fig. 3-38. Any jurisdiction that still allows its installation, I understand, demands grounding; this does not include any places where I am licensed.

I consider the best choice for correction of nongrounded systems to be yanking out what is in place and rewiring, where I'm authorized to do so. The next best choice is to run a grounding conductor, adequately

FIGURE 3-32   A porcelain utility box served K&T.

Figure 3-33   A bushing held wire away from metal.

protected, to a reliable and legal ground. But how much time will this save compared to running a new cable?

Local water pipe used to be considered a reliable ground, but this no longer is true, because of widespread installation of nonmetallic plumbing pipes, sometimes deceptively interrupting metal pipe or tubing that otherwise is continuous. Not surprisingly, I consider

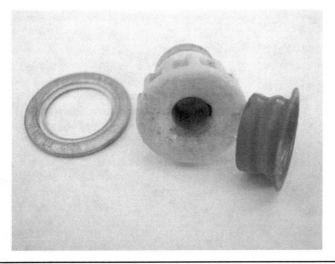

Figure 3-34   This is what went into the bushing.

FIGURE **3-35**   Here's how a bushing looked from the rear.

the most reliable ground to be the grounding terminal in a properly installed loadcenter, specifically, the one from which the circuit originates. I watch out for bootlegs, inadvertent or intentional, including the use of ground wires and bonding where there is no ground. Ignorant people have extended a nongrounding circuit, using a wiring

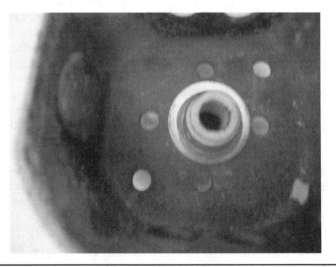

FIGURE **3-36**   Here's how the bushing looked in a box.

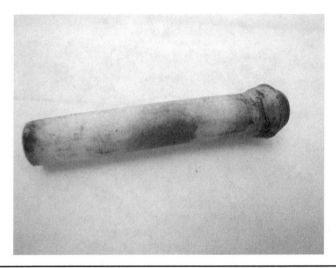

Figure 3-37 This is an old K&T tube.

method that incorporates a grounding conductor, which becomes spurious protection. When the outlet from which they extended the circuit suffers a ground fault, that enclosure and cover plate become energized, if conductive, and the same with every enclosure and cover plate wired downstream. When this has happened, I receive a call about someone getting shocks from their electrical system.

The danger may be even worse if the downstream outlet is three-prong and lacks GFCI protection. If the upstream outlet has a GFCI, and the ground fault is from its line-side energized wire to its enclosure, the GFCI will have no reason to trip, and anything connected to those third prongs will become energized! It will stay energized until something, or someone, draws enough current to operate the overcurrent device. Because of this risk, when I install a GFCI device (as opposed to a GFCI circuit breaker) to permit me to use three-prong receptacles on an ungrounded circuit, I take an extra step: besides applying the "No ground present" sticker, I tape over the line-side energized terminal after attaching the wire, even though I don't do this elsewhere. I see no reason to replace a nongrounding receptacle with a similar two-prong nongrounding receptacle, because the GFCI option is available. Two-prong receptacles may inspire the use of three-prong adapters, particularly dangerous where no ground is present unless there is a GFCI to offer partial protection.

Unfortunately, I've never seen one of the "no ground present" warning stickers on a device or cover, so, I suspect those that are installed don't last long. I also never see a "GFCI-protected" sticker, so nowadays when something goes dead, even in a location that does

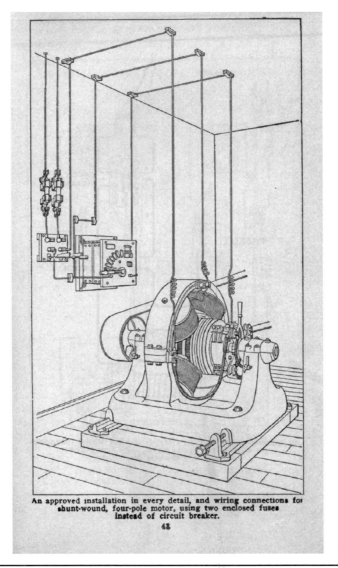

An approved installation in every detail, and wiring connections for shunt-wound, four-pole motor, using two enclosed fuses instead of circuit breaker.
43

FIGURE 3-38 In 1917, open wiring was a norm.

not require GFCI protection, I tend to look all over for a tripped GFCI. In systems that were wired fairly recently, I look only at locations that are appropriate. In older wiring—and it doesn't even have to be all that old—anything goes.

This is a fact of life.

CHAPTER 4

Modified Designs and Changed Practices

This chapter addresses the accumulated changes that show their marks in a building's electrical layout. Looking at these problems carries forward the work described in the previous chapter. In describing old designs and equipment, this chapter's job continues through the next few chapters. This one focuses primarily on guessing the varying logics that guided the decades and generations of electricians and other workers.

Mysteries

I start off with a troubleshooting example. I find that much of the same exploration, intuition, and calculation I rely on in troubleshooting are needed when I am called in to make additions to old wiring, as opposed to repairs.

Puzzling Switches

Here is what I am confronted with, presented from the customer's perspective, and in his words. "There's a funny switch in this room that doesn't control anything." Or she might say, "The light in my closet no longer works."

"I want to know what this wire is; is it dangerous? Can I get rid of it?" poses a similar challenge. Happily, in many of the latter cases, one glance will tell me that, by rights, the wire must serve a doorbell or telephone jack, unless someone was terribly creative in material selection, The latter is something I have come across, which is why I always test. If I am confident about the test results, I can offer reasonable assurance to the customer that the wire or cable poses no hazard, even when I am unable to trace where it leads. (This might not be true if it terminates in a hazardous location, but hazardous locations as such have been rare in my work.)

Some Simple Possibilities

What could that funny switch be? It could be any number of things. It could be a switch that was bypassed by a previous electrician, as he converted a receptacle from switched, often split-wired and half-switched, to unswitched. It could be a three-way switch similarly bypassed so that its mate acts like a plain snap switch.

Blanking off its location would have been a better job. Now that I am the person on the spot, it is up to me to make that change, if it's going to happen. That is not required, so I let the customer decide its fate. Sometimes, it's legal to replace the switch with a receptacle.

If the "funny switch" is not one that was intentionally disconnected, what else could be the solution to the mystery? It could be a switch that has failed in either the ON or the OFF position. Then I know what to do: replace it, test the replacement, and that is that! I will find worse cases, though, in older buildings. They represent examples of what happens when an electrical system is monkeyed with, and a building's layout is modified indiscriminately, over the years.

A simple solution presents itself and makes me look awfully smart, when I ask, "Is there any light, or any receptacle outlet—or even just half of a duplex receptacle—that seems to work OFF and ON without rhyme or reason?" and the customer says "Yes." Even the customer can usually check, once I draw the connection for him, whether the mystery switch controls that receptacle or lighting load.

In another simple case, the switch is working, it has power, and is, in fact, controlling a load. However, the light is outside, or on the other side of a stairwell door, so no one gets the idea to put together the switch and the light. Far more bizarre, and more especially associated with the checkered history of older homes, is the possibility that somebody installed a wall between the switch and the outlet it controls. How was anybody to even think that the dead receptacle in Joe's room was connected to the dummy switch in Sally's? Worse yet, and frequently illegal, is the case where a house has been divided into multiple occupancies and a switch in one turns out to control an overlooked outlet in another.

It's more of a challenge where the load is known and the switch operates unreliably, intermittently—without three-way wiring to blame. In those cases, bypassing the switch, or even replacing it, may be the fastest way to confirm the diagnosis.

Similarly, if an outlet has been mysteriously dead, it could well be that the mystery switch controls it, but failed in nonconducting mode so power could not make its way through to the outlet. This, too, is easy to check.

FIGURE 4-1
Extension cord
wiring is common.

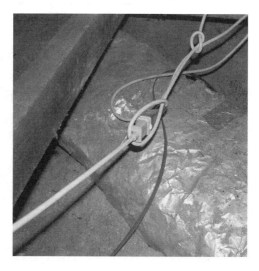

Some Messier Possibilities

Those situations are relatively easy to identify. The next couple of possibilities are real messes. The lighting outlet could well have been buried by someone ignorant of the Code or indifferent to it. I've had customers, or their builders, bury outlet boxes after I refused to do so and I had even explained both the rule and the reason for it. The alternatives I proposed, a blank cover or, where this just won't do, rerunning a line, were considered simply unacceptable. Sometimes when a line needs to be run or rerun, laypersons do it themselves, without involving anyone competent. Examples of this are shown in Fig. 4-1, where extension cords were run, and in Fig. 4-2, where a cord was connected to NM. Figure 4-3 shows the kind of splicing this has included. At least in Fig. 4-4 the plug-connected NM was easy to locate. Even when they ran NM all the way, Fig. 4-5 shows an amateur idea of a home run. Figure 4-6 shows an unwrapped amateur splice; Fig. 4-7 shows an attempt to bond old NM to a box-support bar. In contrast, Fig. 4-8 shows carefully done, carefully taped running splices; Fig. 4-9 shows the same concept in taps.

FIGURE 4-2   Cord connections to NM with plugs are also common.

Figure 4-3   Cord connections even lie buried in insulation.

Sometimes the original fixture was a ceiling outlet, whereas other times it was a sconce. The next-to-worst case I've run into was where the outlet location was right where a wall was added, and the box was buried with a vengeance. The very worst is where the outlet or outlets ended up in another occupancy. I've seen this more than once.

How do I find a buried box? It can be tough. Where I notice a suspended ceiling, I am alert to the possibility that a light was "buried" there. When I see a suspicious dimple in wall or ceiling, as can be seen in Fig. 4-10, that is worth investigating (see Fig. 4-11). If there is no outlet at the center of the ceiling, I give that location a very close look. In these cases, before cumulative damage to my hearing reduced my ability to use the method, I used to tap against the wall or ceiling to listen for suspicious echoes. One engineer I respect, Met Labs's Bruce Moore, used thermography to locate buried connections in an energized circuit.

Figure 4-4   Amateurs can misunderstand supporting and splicing.

FIGURE 4-5
Retrofitters often misapply cables.

FIGURE 4-6 Some amateur splices are quite dangerous.

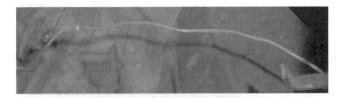

FIGURE 4-7 Many amateur grounding ideas are awful.

FIGURE 4-8 K&T used running splices, soldered and taped.

FIGURE 4-9 K&T created taps the same way.

FIGURE 4-10   Ceiling repairs can point to buried boxes.

Circuit tracers, and even a stud finder, can sometimes help me track cables, and thus locate buried terminations. Unfortunately, in some types of construction, I have found a circuit tracer of little help. I am thinking of a concrete ceiling I encountered some time back; it contained multiple runs of EMT running close to each other. There are also the meaningless readings I got through doubled 5/8 in. drywall under 2-by-12 joists. It may be that the latest tracers are enough better to be more consistently useful for this purpose.

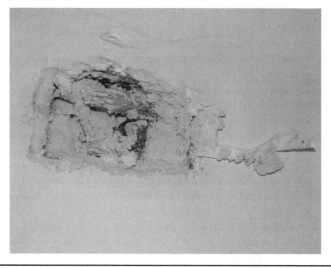

FIGURE 4-11   Buried boxes often lack proper covers.

When I do find a buried outlet, I normally expose it. A buried cable end with no box is even more of a problem. I've terminated the cable in a box and blanked it off, with two exceptions. One is where the cable let me install an outlet. The other is where I could remove the cable entirely. If the cable was embedded in masonry, though, I just made sure that I had killed power to it and called it a day. If my customer refuses to authorize the work, I simply document my recommendation and let it go.

A Buried Switch   Although I haven't seen this often, a dead outlet can mean a buried switch. My guess tends to be that the switch was dismissed as not doing anything and someone removed the switch and plastered over the box.

Somewhat more commonly, a multigang switch box has a blanked section, or a disconnected switch, formerly controlling the outlet that has me puzzled. Other times, the answer to a puzzle is that the switch I am looking for was not disconnected, but given a new purpose. When I have had reason to believe that another, switched, outlet is a relatively new addition, I have sometimes found that the switch controlling the new outlet is the self-same one that used to control the formerly switched outlet I am troubleshooting. In my experience, none of this is very common.

Problems Created by Ignorance About Boxes and Wires

The solution to the case of a removed switch can be as simple as replacing the missing switch. Rarely have I been that lucky. More often, a device has been removed because its space was needed for some other purpose. In such cases, I have chosen among several options. These options also are worth considering on jobs where I am not there to troubleshoot, but rather need room in the box, because I have been called in explicitly to add an outlet or switch at that location.

The simplest solutions entail some compromise of function, or rely on the use of combination devices. When it was acceptable to my customer that the wall sconce and the ceiling light came on together, I have eliminated the need for one switch. Other times, I have installed a combination device, sharing a yoke, when doing so was acceptable to my customer.

Finally, if unit switching is legal for the location and acceptable to the customer, this too can free up a switch for another purpose. Given NEC requirements and intentions, I warn the customer who urges me to put a paddle fan and light on one switch. It is likely that, on occasion, the customer will want the fan on and the light off, and thus will use the unit light switch, on the fixture itself. If the fan/light combination is the sole switch-controlled light in the room, the next user of the room may have to fumble across a dark room to turn on the light. Still, I warn, but I don't refuse to perform the work as the customer chooses.

Ways That I Add Room to Enclosures

Rather than compromise or consolidate, further crowding a box that may not be terribly roomy, I have expanded. When dealing with a sectional switch box, I used to cut open the wall and, with difficulty, add another section. Nowadays, I find volume expansion easier, with at least two manufacturers offering multigang, retrofittable boxes that secure to a stud (and I do mean secure). A less attractive approach is to add a surface extension, using a wiremold box, perhaps multigang.

If the box in the wall does not permit me to add sections, for instance if it is a square box, I am not stuck. I may be able to add a second box adjacent to it, using a chase nipple, or using a short nipple with locknuts for spacing. When I use this option, I'm careful about spacing the new and old enclosures, both with respect to Code restrictions and to how the covers will line up.

Sometimes I've found that an outlet is dead, but its enclosure contains both energized and neutral conductors. A switch leg—and only a switch leg, with no third wire to carry a neutral conductor along—leaves for points unknown and does not carry return current. In some cases, when I simply haven't managed to find out where the switch leg went, I've eventually disconnected it and capped the wires. Where the load needed a switch, I simply ran a new line to it.

Grandfathering Issues

It is time to shift topics back to a different type of decision. I have talked a little about deterioration and about careless installations. However, in old buildings, I frequently am up against at least as many problems in addressing changes in doctrine, in how things are done. These may well present at least as much difficulty as does bad or damaged work. Not all of those doctrinal changes are reflected in NEC changes. I discuss this further in Chapter 10, "Inspection Issues."

Backstabbed Devices

Since the 1993 NEC was adopted, push-in wiring of receptacles and switches became legal to perform only with 14 AWG solid copper conductors. Listed splicing devices that rely on the same design remain legal. Still, I have learned not to use or recommend them, because the installations have historically been less reliable. Cheap receptacles and switches sometimes have offered nothing but backstabbed connections, as in Fig. 4-12 (whose front is Fig. 4-13, its configuration suggesting the vintage). Figure 4-14 is another example in a receptacle, Fig. 4-15 in a SPST, and Fig. 4-16 in a three-way switch, all cut-cost

FIGURE 4-12   This 8B/device cover was backstab-only.

devices. Even if a device like that in Fig. 4-17 were still functional, and still demonstrated sufficient blade retention strength if I bothered checking it with my tension tester (Fig. 4-18), I would replace it when it was time to pigtail the wires, despite it's actually having a screw terminal on each side—it lacks a grounding screw.

On occasion, an ignorant installer will mistake a pinch-plate terminal, which may have a hole for the conductor, for a backstab. I had to red-tag an installer—working under a license!—who just wouldn't

FIGURE 4-13   The 8B cover incorporated three receptacles.

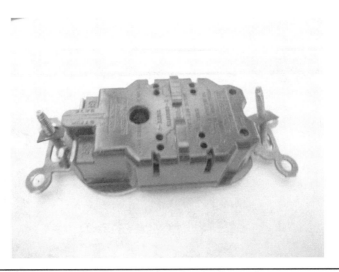

FIGURE 4-14 Backstab-only receptacles often had grounding screws.

FIGURE 4-15 Backstab-only switches were ungrounded.

FIGURE 4-16 Backstab-only three-ways were also made.

93

FIGURE 4-17 Cheap work + worn wires calls for upgrading.

FIGURE 4-18 A tension tester shows a receptacle is shot—or not.

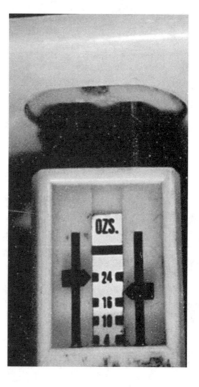

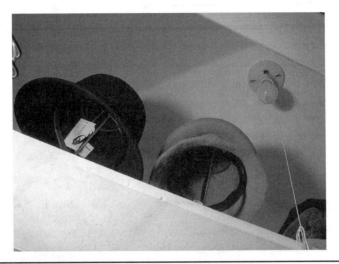

Figure 4-19   Bare lamps in clothes closets used to be legal.

believe me when I told him the GFCIs he had installed were not simple pushwire types, but required screw tightening. I've been fooled
myself by an initial quick glance at some designs, such as back-insert
devices using screw-tightened pressure-plates, a design that certainly
is superior to backstabs.

Closet Lights

Pendants were legal in all closets into the 1930s. Other open incandescent fixtures with a minimum 18 in horizontal clearance from storage
areas, floor-to-ceiling, were legal until adoption of the 1990 NEC. This
is something I have to remind myself about: existing lights in closets, even shiny open incandescent lampholders, may or may not have
been legal to install. They were in violation only where a clothes closet
was relatively shallow. Figure 4-19 shows one that looks too new for its
location. These installations are not readily grandfathered. I wouldn't
replace or rebuild a luminaire located where its installation violates
Code and, if I were asked to approve one on an inspection, I'd be inclined to say no. Certainly Fig. 4-20 shows new-enough wiring that
I doubt the cedar's a retrofit, at least not with the present outlet box
and cable.

Solutions to the Problem of Closet Lighting

Although an otherwise unacceptable incandescent lampholder or luminaire does not become acceptable by use of a screw-in CFL or LED,
replacement of the luminaire with one that accepts only a fluorescent
or LED is fine, as long as the location complies with the easier spacing restrictions. If the location forbids even this, I like to blank off the

FIGURE 4-20 I wouldn't grandfather a refitted clothes closet's light.

outlet. This reduces the temptation for some amateur to get the fixture working, possibly with even riskier, high-resistance splices and terminations. A tidy job of blanking it off also reduces the likelihood that someone will remove the fixture and plaster over the outlet.

Here are some additional solutions I've suggested to customers who needed light in a closet. A cheap, inelegant, low-technology answer for occasional-use clothes closets is a battery-operated light stuck to the wall. Sure, batteries die. Sure, this answer is not pretty. But if the customer just does not have the money, or does not want to spend the money, to hire me to do something more sophisticated—probably involving a permit—it is a simple fix.

The next solution for those clothes closets that are shaped so as to accommodate a luminaire within the Code's restrictions, is to replace the bare-lamp fixture with an enclosed incandescent, or perhaps a fluorescent. In some jurisdictions, such a straightforward swap doesn't even require a permit. If the wiring is in decent condition, the swap takes very little time.

The remaining solutions involve relocating the fixture. Many times, but not always, customers are far less concerned about the looks of what I offer to run on a closet ceiling than they would be if it were out in the open. This means that the solution has been simply putting a surface extension over the outlet box, adding a blank cover, and running cable from the extension box to a fluorescent strip over the closet door. A slightly less unsightly approach I've used is to do the same thing with surface raceway.

Where the closet dimensions are such that even this is not legal, I put a light outside, shining in. Usually, I can work it to run the power from inside the closet, even though the outlet will be in the ceiling outside. This means that often I can avoid creating an offensive mess in the actual bedroom ceiling that might require professional wall or ceiling repair.

I have not yet installed an LED-containing clothes rod, although the idea is intriguing.

Grandfathering: The Bigger Picture

Although I generally prefer to keep a subject to one chapter, it is worth looking at grandfathering in a number of places. This is its second appearance before Chapter 11, Grandfathering, Dating, and Historic Buildings. Various Code-enforcement issues are involved in grandfathering.

There are a number of criteria to consider:

- Was this installation that I have been called to repair ever legal, unlike what you see in Fig. 4-21?
- Is it dangerously deteriorated?
- Is it palpably unsafe for some other reason?
- How much change is needed to make it legal?

FIGURE 4-21   Eaves often hold illegal security lights.

- How much work am I going to perform on it and to how great an extent will the installation be modified?

- Does the NEC explicitly require updating the installation, as is the case with GFCI protection?

My Responsibility with Regard to What Already Exists

Breaking the questions down, if an installation never was legal and never will be legal, my getting it to work might land me with responsibility for the whole mess, in the event that anything ever goes wrong (depending on the lawyers). If an installation was otherwise legal, but clearly installed without a permit, and installed recently, I won't add onto it unless the customer is willing for me to pull a permit that notes, "Work begun by another." It used to be that this would result in a transitional inspection. In cases where I do work on an installation containing work that had been begun illegally, my invoices and records specify just what I've installed, as opposed to finding work already in place, documenting the situation for the AHJ.

It is normally not my responsibility as a contractor to engage in historical research on a building, to determine whether something preexisting originally came through inspection. Here are two questions I ask. Yeses get my go ahead.

- Can a reasonable person take the installation as having been professionally installed—perhaps, admittedly, by a competent, but sloppy, professional? Figure 4-22 shows an example that might have passed inspection, but if I'm working there, I

Figure 4-22   Code calls for repair around *box* edges.

will add spackle to close the wall around the boxes, not merely the yoke ears.

- Is it reasonable to assume that the original installer pulled a permit and passed inspection—although maybe a very hurried inspection?

Example: Overfusing   Some installations I encounter are technically illegal, but really don't strike me as being so bad. An old, grandfathered fusebox (meaning, in this case, that it was installed before adoption of the 1923 NEC) may fuse both energized and neutral conductors. It could very well now have a 30-amp Edison-base plug fuse in series with 14 AWG branch circuit conductors on its neutral, to ensure that the fuse on the energized conductor is the one to blow. Not a bad idea. However, in such cases I strongly discourage customers from leaving extra 30-amp fuses around, so they do not succumb to the temptation to use those as the energized conductors' protection, not even temporarily, when they blow fuses protecting *those* conductors.

Of course, grandfathered or not, if the old fusebox is service equipment, with a main bonding jumper, and there is sufficient legal room on the ground bar, with the customer's permission I'll move all those neutral wires into spare terminals on that bar. No holes available on it for more conductors? If the cabinet was designed to accommodate an additional grounding bar and there was space to do so, I've added one. Otherwise, if there was room to do so, sometimes I've spliced branch circuits' grounding wires, and run one fat pigtail from them to the bar.

That's not all I do about fuses. If there is any evidence that the customer has overfused, I take it as my responsibility to install Fustat adapters, with one exception to this rule. When the fuseholders are shot, or if any of the internal connections in the loadcenter appear to be damaged or otherwise likely to add significant resistance, I tell the customer they cannot be used. Usually, but not always, this means that the loadcenter needs to be replaced. To have been installed legally, loadcenters with Edison-base fuseholders have to have been installed before the jurisdiction adopted the 1941 NEC.

Example   One jurisdiction where I have worked both as a contractor and as a Third-Party Inspector (TPI) is Washington, D.C., which still enforces the 1996 NEC, as I finish preparing this edition, in early 2010. I haven't checked the adoption dates of earlier versions. What's grandfathered in D.C., in fact what's presently legal to install in D.C., differs considerably from what's legal a mile outside the city limits, in Maryland or Virginia. I would not be surprised to learn that this was true in the 1940s and 1950s as well.

The other situation where I used to leave out the no-tamper adapter to prevent overfusing was when the customer has installed a minibreaker, a magnetic-trip breaker in the shape of a fuse. This is relatively unlikely to be unwittingly replaced with an oversized fuse. However, minibreakers are not Listed for branch-circuit protection, but only as supplementary protective devices. It was a long time till this was pointed out to me.

An Esoteric Switching Layout: The Carter System

I will talk about one final common design that has changed and has not been used for quite a long time. However, the issue of Carter System switching is so large that I have left it in Chapter 5, which follows.

CHAPTER 5

An Esoteric Switching Layout

This chapter is devoted to variations on one old-wiring layout, long-illegal, but also long-employed. I have not updated the basic Carter System diagrams from the first edition to reflect modern terms, or the comparison diagram of our present-day system.

Carter System or "Lazy Neutral" switching goes by many names. Creighton Schwan heard, "San Francisco Three-way" in Los Angeles, and "Los Angeles Three-way" in San Francisco. Andy Cartal knows it as "hot trolley" switching. Call it what you will, it is an example of something that went from initially being commonplace and legal to many decades of being commonplace, and illegal, because it means switching the neutral. At present, a small minority of electricians would recognize it. It certainly was quite unfamiliar to me when I was newly trained. When I first untangled an example, it looked like the most outrageous screwup. Yet, at one time it was a very attractive approach, because it offered an awfully clever solution to the challenge of doing the most with the least—kind of like running multiwire circuits from a panel. It certainly was used for decades past the time when it became illegal.

Modern style three-way switching, originally known as two-way switching, was known far back, as shown by Fig. 5-1, "reproduce freely" instructions from an old three-way, made by a defunct company. Showing both standard and Carter Style switching, it is so old that it antecedes the convention of switching the energized conductor and not the neutral. The NEC's 1897 source document requires a circuit's conductors to be run together when in conduit. By 1915, if not earlier, the NEC required that all wires of a circuit run in armored cable to be in the same cable. While the 1917 handbook from which Fig. 5-2 was drawn likewise does not always acknowledge our concept of switching only the energized conductor, it seems to do so sometimes, as can be seen in Fig. 5-3. However, it doesn't even describe Carter System wiring, much less provide a diagram of the approach.

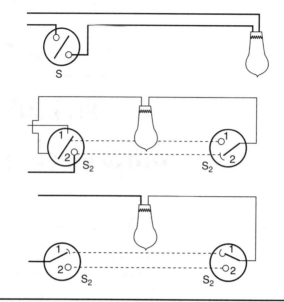

FIGURE 5-1 Carter switching once seemed equally okay.

WIRING DIAGRAMS FOR FLUSH SWITCHES

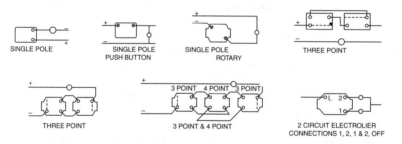

SINGLE POLE

SINGLE POLE
PUSH BUTTON

SINGLE POLE
ROTARY

THREE POINT

THREE POINT

3 POINT & 4 POINT

2 CIRCUIT ELECTROLIER
CONNECTIONS 1, 2, 1 & 2, OFF

FIGURE 5-2 In 1917, switching neutrals still seemed okay.

WIRING DIAGRAMS FOR SURFACE SWITCHES

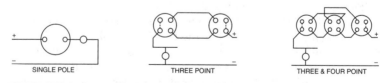

SINGLE POLE

THREE POINT

THREE & FOUR POINT

FIGURE 5-3 Some old diagrams don't show Carter System designs.

The Logic of the Design—What Makes the Lights Light

Carter System switching allowed electricians to do several things using two conductors rather than three, but there was a price. The largest price was tied into one fact, which I'll describe in terms of a screw-shell lampholder. In normal practice (Fig. 5-4), the shell is permanently connected to a neutral and the center contact to a switch leg, consisting of an energized conductor coming ultimately from one direction.

Carter System switching functions differently. So far, I've only encountered examples used for lighting control. When working, they switch both energized and neutral conductors and send a switch leg to the switch-controlled outlet from each switch: one to the shell and the other to the center contact. The essential, simplest, purest layout requires just one conductor from each three way—although I've never found it this simple in the field. The way in which this functions is shown in Fig. 5-5. Assuming the lamp and fixture are in good working order, sometimes the light is on and receiving power just the way it would, wired today: the center contact of a screw-in lampholder, or its equivalent in other lights, is connected to an energized conductor, and the shell is connected to a neutral. At other times, the light shines but it is fed backward: the shell is connected to an energized conductor, and the center contact is connected to a neutral.

When the lamp is dark, sometimes no power is going to the lamp, because each contact is connected to the neutral. However, sometimes the lamp does not light because energized conductors are connected to both the shell and the center contact! It is energized, but lacks a return to complete the circuit, as shown more graphically in Fig. 5-6. In short, one price the system always exacts is that lampholder polarity goes out the window.

I'll restate this with reference to the terminals of a three-way switch. In the Carter System, one TRAVELER terminal at each switch is fed by the energized conductor, while the other is fed by the neutral. Thus, depending on handle position, each switch can connect its

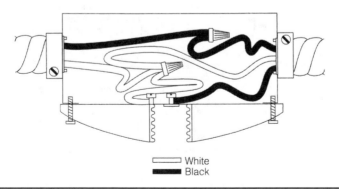

⊐ White
■ Black

FIGURE 5-4 A neutral splice at the light says "not Carter System."

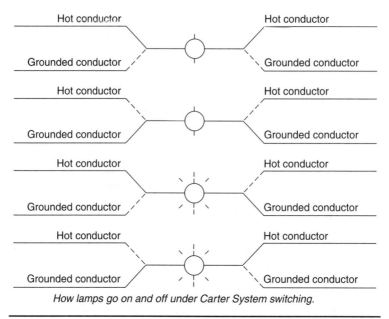

How lamps go on and off under Carter System switching.

FIGURE 5-5

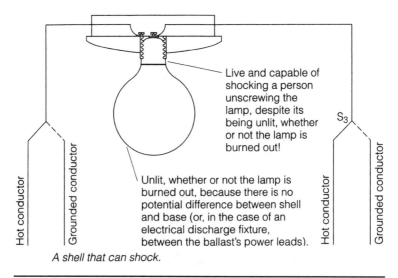

Live and capable of shocking a person unscrewing the lamp, despite its being unlit, whether or not the lamp is burned out!

Unlit, whether or not the lamp is burned out, because there is no potential difference between shell and base (or, in the case of an electrical discharge fixture, between the ballast's power leads).

A shell that can shock.

FIGURE 5-6

COMMON terminal to an energized conductor or to a neutral. At the switched lampholder, the conductor coming from one switch's COMMON terminal is connected to the shell and the conductor coming from the other switch's COMMON terminal is connected to the lampholder's center contact. I also find receptacles as part of wall sconces; here, the polarity of the blade slots and any conductor color-coding, are meaningless or misleading. At any moment either the silvery or the coppery screws may be connected to the energized conductors or to the neutrals.

Multipoint Switching and the Carter System

Describing the operation is not the same as explaining how pipe was wired or how cable was run.

Comparing Carter Designs With Standard Multipoint Switching

To highlight the differences, consider how multipoint switching always is arranged now when legally wired, shown in Fig. 5-7. The energized wire coming from upstream feeds power into the COMMON

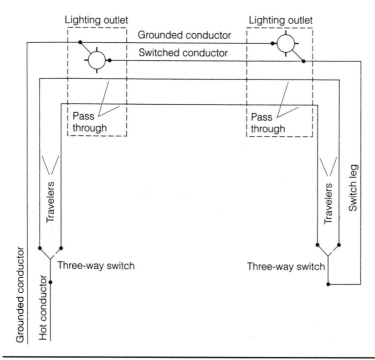

FIGURE 5-7 Only modern three-ways need multiple travelers.

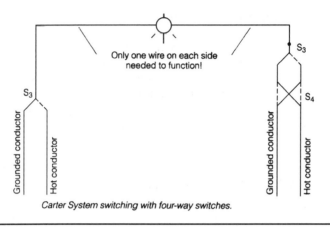

Carter System switching with four-way switches.

FIGURE 5-8

terminal of a three-way switch. The switch directs the power to one of two TRAVELER terminals. The second switch is fed by conductors coming from those TRAVELERs. At any moment, one of the two TRAVELERs is connected to the COMMON terminal of that second switch, and through it to the switched outlet. The neutral connects to the load side of that switched outlet, uninterrupted. This should be familiar. It is intended to light when the center contact, or equivalent, is connected to the energized conductor, the only one that's switched, which ultimately reaches it through only one incoming cable or raceway. The fixture is intended to remain unlit when the center contact, or equivalent, is disconnected from the energized conductor, and, in fact, no longer is directly connected to any conductor.

More Complicated Carter System Designs

I don't remember coming across a Carter System installation that employs more than two switches to control an outlet, but I might have, at some point. I don't doubt that they were installed, as shown in Fig. 5-8. An energized conductor and a neutral would feed into a four-way switch; that switch would use two wires to daisy-chain in series with any other four-way switches; the last four-way switch would then feed the TRAVELER terminals of a first three-way switch. A second three-way switch would be fed by an energized conductor and a neutral, either directly from a source of power, or, similarly, through one or more four-way switches. The three-way switches would then connect to the outlet as described earlier.

Recognizing Carter System Switching

An old rule says that in troubleshooting, it almost always is wise to start the investigation from the source of power. I follow it, loosely.

With switching systems, initially I don't know whether the furthest upstream local point on the circuit is located at the lighting outlet or at the switches themselves. As I mention in my discussion of troubleshooting, it often makes sense to me to start at the switches. I have two reasons for this. First, since they have moving parts, switches can be expected to wear out. Second, they are usually easier to access than the lights—no ladder is needed and I tend to feel less at risk of causing damage.

Looking Inside the Switch Boxes

Carter System switching was employed most commonly where several multipoint switches were ganged together. The reason is that the installer gained greatly in efficiency by reducing the number of TRAVELERs that otherwise would be needed. Besides, what's the point in running a single conductor to the light? Sometimes the lighting outlet was used as a junction box, with a second Carter System type switch leg or two traveling from it to a second, independently controlled lighting outlet, as in Fig. 5-9.

Rarely is it hard to identify the presence of Carter System switching at a multigang switch box. I notice that several three-way switches have two of their three terminals, the TRAVELERs, wired together. I find one of two things. Either there will be two splices, each of which contains one wire coming from each of the three-way switches, or a wire will be daisy-chained from each of two terminals of one to two terminals of the next. When I can examine a switch easily, I look for markings, or for color differences in its terminal screws. If only one of the three terminals is not connected to the other switches, that's enough; but if it is marked COMMON, or has a different finish, this proves the case. The markings on the terminals of these ganged switches usually are not very accessible, but I don't find them essential.

With standard three-way switching, in contrast, I might find one, and only one, conductor, the COMMON, connecting switch to switch within a box. Grounding/bonding terminals, of course, have nothing to do with circuit logic. However, not only do these conductors, when present, not tend to be daisy-chained, but they tend to be bare. Also, I tend not to find grounding terminals on switches in wiring of this vintage, unless they are replacements.

More often than I'd like, the boxes are very crowded, the terminals are taped, or the wires are very short and hard to get at. As a result, I'm not clear about where all the wires go. A mistake in this can cost an awful lot of time. In part, the difficulty is increased because I recognize that insulation and even wire may be fragile and because I may have been mistaken about which, if any, are energized. I treat the wires accordingly. It also can be quite hard to read the markings on switches, either because they are marked with faint embossing, or because they

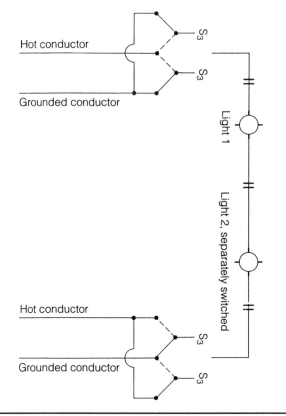

Hot conductor

Grounded conductor

S₃

S₃

Light 1

Light 2, separately switched

Hot conductor

Grounded conductor

S₃

S₃

FIGURE 5-9 Ganged Carter System switches need more switch legs.

are not accessible without pulling the switches all the way out. If I'm sure that at least one terminal on each switch is connected to the others, but I cannot readily see from markings on the switches which terminal is which, they could be COMMONs, connected to the energized conductor. I either pull out the switch or, with energized circuits, test with my voltmeter.

Here is another ambivalent case. Suppose I find one three-way switch, in a single-gang enclosure. It is fed by at least two cables or conduits, so I know that it contains more than just a switch leg. The conductors connected to two of the switch terminals come from splices (and I don't mean just pigtails installed to lengthen short conductors). In this case as well, I strongly suspect the presence of Carter System switching (Fig. 5-10).

This is how I check. If circumstances are such that I feel safe testing with power on, I remove the lamp from the lampholder; this avoids backfeeds—if and only if (IFF) it's the only load controlled. Now I confirm that one of the splices feeding the switch, or one of its TRAVELER

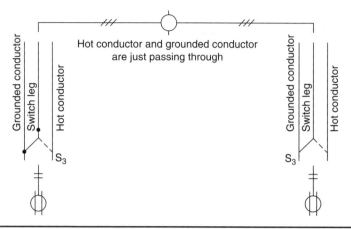

Hot conductor and grounded conductor
are just passing through

FIGURE 5-10   Extra wires that aren't switch legs can cause confusion.

terminals fed from a splice, is energized, and that its companion splice or TRAVELER terminal is roughly at ground potential. I confirm that the positions of the switch handles, both there or across the room, do not affect these readings. I have Carter System wiring, unless it might be that the wiring is standard. In this case, the switch is defective, unable to disconnect the TRAVELER from the energized COMMON. Furthermore, there's either a stray ground or a second, occupied lampholder providing a return path. I can confirm the presence of that return path up at the light.

Relevant History
Back in the days of rag wiring, there was a version of ungrounded NM composed of two rubber-and-cloth insulated wires within a sheath of impregnated cloth. However, there was also a distinct assembly of two rubber-and-cloth insulated wires within a sheath of nonimpregnated cloth; this was designed to be pulled into conduit. W. Creighton Schwan, who knew so much about the history of electrical materials and the electrical code, told me that it was known as "duplex."

Example   Sometimes, I'm not the first person to troubleshoot the system. This can complicate matters. Here's a case where, at first, I formed the incorrect idea that the switches were wired Carter style.

Taking over repairs on a 1923 house, I was initially misled by one of these "duplex" assemblies. Four three-way switches had been removed before I was called in, two in a box in the upstairs hall and two downstairs. Any splices had been undone. The two conduits running from switch box to switch box each contained duplex plus a single conductor, and, until I took a more careful look inside the boxes, I did not realize that these three entered the box from the same conduit. As a result, I did not realize, at first, that the three-way switches communicated with each other via three wires, as opposed to two, because two of the wires came into the enclosure together; this was the duplex cable. Happily (by which I mean that the arrangement made intuitive sense), in each case the duplex conductors were used for the travelers and the singleton for the common.

When I'm troubleshooting, my identification of what design the wiring system employs sometimes is uncertain at first; misinterpretation of my meter readings misleads me. The energized or approximately ground-level reading that should be present at one of the terminals is missing because of a bad connection somewhere; other times, the normally energized conductor is near-grounded through a load, or thanks to a ground fault. Fortunately, when lighting systems go bad, switches are the most common culprit. Therefore, in most cases, I find energized conductors and neutrals present in most Carter System switch boxes.

When uncertain about a switch, I've killed power, disconnected the leads, and tested the switch in isolation, using my continuity tester or ohmmeter. At least, then, I know whether the switch is the source of the functional problem. On relatively rare occasions, I trace the problem to a bad splice or a broken or ground-faulted conductor. When I find a three-way switch with one energized terminal, apparently a TRAVELER, and the other apparent TRAVELER at approximately ground potential, this can be easy. If the third terminal, seemingly the COMMON, does not alternate between those two polarities as I flip the handle, I most likely have a defective switch, although I don't necessarily replace it. There is the possibility, for example, that it is wired Carter style, the third actually is the COMMON, and it's being energized through the load. (Read through the discussion of fixes, a little further down.) The same reasoning applies when the seeming COMMON acts like a neutral regardless of switch position.

If one switch tests out okay, I may need to try its partner. Just as is true with normally wired three-way switches, in Carter System wiring there are several failure modes. If one switch continues to conduct but stops switching, the other will be constructively turned into a single-pole, single-throw switch—possibly working upside down, possibly switching the neutral rather than the energized conductor, but working. Because people often rely on only one switch of a pair that work together, sometimes, when the defective or dying switch is the unused or less-used one, the problem hasn't been noticed until the second switch also failed. I return to this point a little further on.

The other failure modes are simpler. The other usual failure mode I've found in three-way switches is to stop conducting altogether through at least one internal contact. If the disconnected contact is the COMMON, the light may stay off until the switch is replaced—whether wired Carter style or normally. The third type of failure is intermittent function because of either a switch failure or a loose connection. The fourth failure mode is one that I've rarely found: there is a short between the switch's two TRAVELER terminals. With Carter style-wired switches, this type of failure is a hot-to-neutral short, sure to trip the circuit's overcurrent device if its wiring is otherwise functional and the short is solid enough.

Troubleshooting at the Switched Outlet

When both switches, and the wires attached to them, test out okay, I've next looked at the light or lights. As mentioned in the troubleshooting chapter (Chapter 3), I've started there for other reasons as well. Here's how I've identified Carter System switching from tests at the lighting outlet.

Sometimes I've even been able to do so without taking it apart. I've found the lamp did not light because it was not making good contact—corrosion was clearly visible. My tester scraped through it, informing me that power was making its way up to the lighting outlet. Without taking anything apart, when I've found both shell and center contact energized, or both at ground potential, I've found fairly strong evidence of Carter System switching. I've found far weaker evidence when their polarity was reversed, with shell energized and center contact at ground potential. Finding normal polarity does not, however, prove the absence of Carter System wiring. Flipping the switches and retesting at the light gives me further necessary information.

There are other possible explanations for each of those findings. Reversed polarity has simply been the result of someone's ignorance, or haste, or of someone misreading the color-coding of very faded insulation and not checking with a voltmeter. It is not so rare for the white conductor to be energized and the black to be neutral, even without faded colors. With a normal switch layout, sometimes when both contacts are energized, the reason is that the neutral is broken somewhere downstream along the circuit and another load is backfeeding to the light. Usually, when both test out as being at roughly ground potential with the lamp removed, this is because Carter style wiring was, in fact, used. However, in other cases, because of ground faults, wires feeding a lampholder from a normal switch test out as though they were both neutrals.

Going into the enclosure behind the luminaire does not automatically, necessarily tell me more about whether I have Carter System switching. If the local source of power does not come up through the switches, but enters the room or area right there in the lighting outlet box, this box should contain a neutral. Finding this connected to the fixture shell is proof that what I am seeing definitely is not Carter System switching. Even where this splice feeds a lampholder's center contact, I have simply found reversed polarity. This is still a different, and simpler-to-fix, situation than Carter System switching. In most cases where power enters from the switch, especially when the energized or neutral conductor is interrupted, the lighting outlet does not tell me if I am seeing the Carter System.

One setup causes some confusion as to whether there is standard or Carter System switching, because the lighting outlets looked the same. This is where the switches simultaneously control multiple lighting outlets. In this situation, the lighting outlet feeds both lampholder terminals from splices. From those splices, conductors continue on to the other lighting outlets. Contrast paralleled lights controlled by modern three-way switching with paralleled lights controlled by Carter System switching. In each case, two conductors have to travel from light to light, but this is as far as the parallel goes.

There are several other cases where the system in use is not apparent. Where conduits or cables that contained several conductors enter the lighting outlet and the lampholders are not fed from splices, I have to test at the switches.

A Rough Parallel to Carter System Wiring

An illegal, and even more dangerous, parallel to Carter System wiring is the use of a (bootleg) ground return.

What Carter System Switching Accomplished

One problem that Carter System switching solved, and solved very nicely, originated from the transition from gas lighting to electric. A room might have had one light burning—literally—in the center of the ceiling, fed by gas pipes that later were used for wiring. In consequence, an old-time electrician found a pull-chain luminaire in the ceiling, or none at all, and that's all the power there was. The family's clothes iron would be plugged into it, probably through an adapter. Now he was called in to modernize by installing receptacles in two of the baseboards. The customer also wanted to control that light, presently operated by a pull chain (or, if there was no light, wanted to install a new light and control it), from each of two room entrances. Of course, the electrician preferred to economize by using the least cable possible. Carter System wiring allowed the electrician to do the job using less cable and one less conductor in each run (Fig. 5-10).

Here is another problem the system solved. That old-timer found a switch-controlled light, with power—both energized conductors and neutrals—coming through the switch box. He wanted to add control from a second switch. He preferred not to have to rerun a three-conductor cable from the existing switch to the light, or around from a second switch on the other side of the room to the first switch. Carter System switching accomplished this using one less conductor in each cable, which meant less box crowding. He replaced the single-gang switch fed from the energized conductor with a three-way switch fed by both. He still had to run two wires from the new switch (presuming that it is located at a source of power), but using modern-style three-way switching he would have needed three.

A Closer Look at the Carter System's Drawbacks

Before considering fixes for malfunctioning Carter System installations, I need to keep several drawbacks in mind, in cases where I might be considering treating the setup as grandfathered.

Polarity Loss

The loss of reliable polarity is a significant drawback to Carter System switching. Very frequently, I find only one of two three-way switches is regularly used. This means that users know which handle orientation indicates power to the light is off. With the Carter System, unscrewing a burned-out lamp that seems to be turned off can put the user in contact with an energized shell.

It doesn't take Carter System wiring to do this; even a single-pole switch that is wired into the neutral rather than the energized conductor poses this risk. In that case as well, color-coding can't be relied on to ensure that a white wire reaching the lampholder from either switch is a neutral. This fact can increase the risk to future workers as well as to end-users.

Induced Current

There is another cost besides the loss of polarity and color-coding. Despite the fact that each cable or conduit contains several conductors, with Carter System wiring only one may be carrying power when a load is fed. The first problem this creates with installing or grandfathering Carter style wiring is that current could be induced, as shown in Fig. 5-11. This could possibly flow across resistance so as to generate heat, whether in cable armor, in the walls of a metallic electrical box, or between armor, cable connector or clamp, and box. It could arc across a small enough gap. Theoretically, it could even result in shock, if the grounding path is flawed. A second, lesser risk is that by inducing current in surrounding metal, this configuration could result in a choke effect on the power wiring, causing an undervoltage. Any such loop certainly can generate massive EMI, as shown in Fig. 5-12. If nothing else, this can play havoc with electronics.

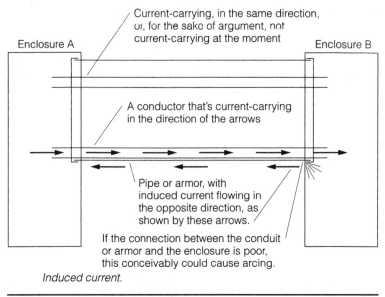

Current-carrying, in the same direction, or, for the sake of argument, not current-carrying at the moment

Enclosure A

Enclosure B

A conductor that's current-carrying in the direction of the arrows

Pipe or armor, with induced current flowing in the opposite direction, as shown by these arrows.

If the connection between the conduit or armor and the enclosure is poor, this conceivably could cause arcing.

Induced current.

FIGURE 5-11

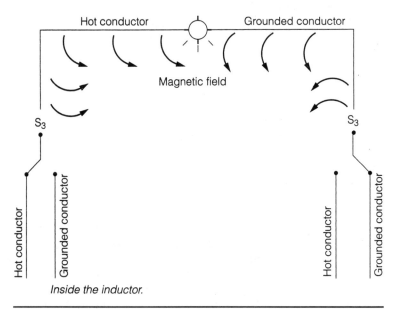

Hot conductor

Grounded conductor

Magnetic field

S_3

S_3

Hot conductor

Grounded conductor

Hot conductor

Grounded conductor

Inside the inductor.

FIGURE 5-12

Responding to the Carter System

My intention is not to provide abstract descriptions, but to talk about what I do in the field. I do not mean to say that I generally have fixed rules about how to treat these situations.

Even If It Is Broke, Must I Fix It?

Before I talk about some solutions, it is worthwhile noting that no AHJ has told me of red-tagging a minor repair to one of these systems, such as replacing a defective switch. I don't know that I would do so, though. I hate to repair Carter System layouts, and don't remember when I have. Usually I convert these to single-point switching. Electronic switching is another option where converting the design to safe three-way switching is difficult.

How inspectors enforce the law can vary a great deal, as I discuss in more detail in Chapter 10, "Inspection Issues." The same inspector who will let contractors replace switches without a permit will not necessarily let them replace luminaires. The circumstances are different.

There are arguments on each side. By one argument, it is okay to replace the switches, but not the fixture. I'm making zero changes in the system's components when I replace a switch with a switch that is functionally identical, except perhaps for grounding. Anybody working on the associated fixture will see that the fixture is old, according to this argument, and, accordingly, will be careful with their assumptions about how it is wired. This sort of careful, "better watch out" thinking is one reason I didn't dismiss the odd strip on the ceiling in Fig. 5-13 as architectural detailing, but checked out my suspicion. Figure 5-14 shows how justified that suspicion was.

By this first type of reasoning, fixture replacement is treated differently. (In a few jurisdictions, all fixture replacements call for permits, but not so device replacements.) Contrary to the situation with

FIGURE **5-13** Zip cord wiring hides easily.

FIGURE 5-14 Zip
cord is pinched
under canopies.

switches, if someone notices a new fixture, and knows or presumes that a professional electrician was the installer, he or she will be more inclined to assume that polarization is normal, and that unscrewing a lamp cannot result in being zapped by the energized shell. This represents one perspective on Carter System repairs.

Some see things the other way around. It also can be argued that it is okay for the fixture to be replaced, because there is nothing wrong in itself with the fixture's wiring. (However, when I install the new luminaire, in most cases these would prevent me from complying with the requirement to connect this lead or contact to a switched, energized conductor and that lead to a neutral.) On the other hand, in some people's views the switching is the source of the problem, and the switches are, clearly, in violation. Therefore, if I replace one switch, I need to correct the whole setup.

Which argument is right? This is a case where there is no right or wrong answer. Experts say, "This is enforcement, not *Code*." When this is the case, it is helpful to step back and look at the bigger picture: any legal Carter System wiring was performed close to a century ago and intrinsically it is relatively unsafe. Most that remains in use, in fact, was not legal at the time of its installation, so it does not warrant grandfathering. This is all a customer should need to know in order to choose to upgrade Carter System switching to something more modern. Still, I usually don't lay down the law.

Fix-It Options

Now it is time to explore how I can modify the system. I consider assorted options.

Elimination

Here is a simple, quick fix: I use one single-pole, single-throw switch. Most rooms with multipoint switching do not really call on both switches, or, with four-way switching, all of the switches. When I eliminate all but one switch, the customer is left with a switch box from which two conductors go up to a lighting outlet. Most of the time, this is fine with them—and legal.

This is how I often do it. Suppose the source of power is at one switch. From there, a three-wire cable brings an energized conductor, a neutral, and a switched conductor up to the light. From the light, energized conductors and neutrals continue on to various other outlets along the circuit. The other switch receives a three-wire cable, which feeds energized conductors and neutrals from the lighting outlet (conductors that the second switch sends along downstream) and allows it to send a third, switched, conductor back to the light.

I want to change this from Carter System switching to single-location control. Eliminating one switch means that, from whichever one the customer decides to keep, three conductors will continue to the lighting outlet, with the switched one going to the lampholder's center contact. The shell now will be connected directly to the neutral. The switch location that the customer decides to dispense with has its switched conductor disconnected and capped at both ends, Sometimes, I'm able to replace the switch or switches I removed with a receptacle, when doing so is legal on this circuit at this location. After all, the enclosure in which a Carter System-wired three-way switch originates must have both energized and neutral conductors available.

Rewiring

In cases where it will be unsafe to eliminate a switch, such as the top and bottom of a staircase, and other cases where elimination is unacceptable to the customer, new wiring usually is needed. On very rare occasion, I am working in a commercial occupancy, and the customer assures me that downstream receptacles that are fed through this wiring can be eliminated. This can free up additional conductors that I can repurpose as TRAVELERs.

Troubleshooting

I am going to focus on the relatively common failure modes. Two simultaneous failures can hit a system, but I haven't seen two of these switches fail simultaneously. This doesn't mean I am not called in to troubleshoot systems where both three-way switches have gone bad. This is far from impossible, and not even terribly unlikely. However, they are exceedingly unlikely to fail simultaneously.

Unused Switches   What happens is this: one of the pair fails, but the users simply don't get around to having the problem fixed. Instead, they start using its partner. Often, over time, they totally forget about the first! There is a straightforward indication, if I'm troubleshooting. Sometimes, a three-way switch will be utilized, as an SPST, for one reason or another. (Inventory engineering is a common one.) When it is, though, it just has a conductor attached to the COMMON terminal and a second to one or the other of the TRAVELERs—not both. Both? It's a three-way, and I want to know where its partner or partners are.

However, I acknowledge that when I haven't had the luxury of hunting and the customer has not minded, I have simply rearranged the conductors at the one switch box that is used, so that an SPST connects the energized conductor to the switch leg. At the light, I reorganize the connections so that the other side of the luminaire is connected to a conductor that remains, I hope permanently, a neutral. The resulting arrangement is legal and safe, if there are no other hazards.

CHAPTER **6**

Unusual and Odd Equipment and Designs

T his chapter covers types of installation that I hadn't seen until
I began moving into this specialty. I talk further about rare
and dated systems in Chapter 7, "Commercial Settings," and
Chapter 4, "Design Changes."

Loadcenters in Unexpected Locations

To begin, mounting a fused cutout in a ceiling once was a very common
practice. This location was chosen when uncluttered wall space was
unavailable. In a room with a low ceilings, or one where a "working
platform" brought users to within 6 ft. 7 in. of any switches or breakers,
the installation would be legal. Anyway, these usually contained plug
fuses, unswitched. Until finally locating a subpanel in a ceiling, I've
had a frustrating and fruitless search elsewhere. I've even figured that
the dead 14 AWG black wire at an outlet was broken somewhere upstream along the circuit, and unfairly presumed that it was protected
solely by the two-pole 40- or 60-amp feeder fuse or circuit breaker.

My encounters with hidden panelboards—which I've found particularly in garage ceilings—are broader than this specific example.
If power is dead at several outlets while all incoming phases or legs
in the main panel are energized, all its overcurrent devices are conducting, and there are unmarked two-pole circuits, I hunt <u>hard</u> for a
subpanel before delving into too many outlets.

First, I eliminate the obvious: in a residence, the unmarked two-
pole circuits often feed electric ranges or dryers, or unused range
or dryer outlets. They've served present or former water heaters or
electric furnaces, air conditioners, or heat pumps. Except for electric furnaces, and heat pumps with resistance backup elements, the

overcurrent devices for such loads generally will have lower ampere ratings than those that feed subpanels. In a multitenant building, residential or commercial, they've fed another occupancy. In any
occupancy, they can feed a subpanel in an outbuilding. In a grocery, they may serve walk-in coolers, rooftop equipment or outdoor
condensers.

If none of these possibilities bears fruit, I look for subpanels where
I would never place one. I look behind art, furniture, cabinets, book-
cases, and other storage. Some of these locations are even legitimate. I
include places where, clearly, no electrical enclosure should lurk, and
certainly no overcurrent devices. Access? Hah! They may not have
been touched for years.

Example   I spent a couple of hours tracking down a subpanel behind
a mirror on the second floor of a rectory, in what was, in all but name,
a different occupancy.

- Could a subpanel be hiding in a bathroom? Yes. Until adoption
  of the 1999 NEC, outside of dwellings and guest rooms this
  was even legal, as long as the location comlied with other
  rules, such as clearance requirements.
- In a location that prevents its cover from opening? Way too
  frequently.
- In the back or side of a clothes closet or kitchen cabinet? I've
  found them there, too.

The most important thing I've learned about these oddball installations is to keep in mind that they exist. I've certainly learned not to
count on a panelboard directory cuing me by saying enough about
the loads that its circuits presently control.

How I Deal with Oddly Located Overcurrent Devices

Once I find a fusebox in an unexpected location, I can replace the
blown fuse. But is that enough? When I discover one of these oddly
located panels, there are three or four more things to do. One is to
warn the customer of its location and explain why it's unsafe. Another
is to create a record of the notification—not just in notes, but on my
invoice. I like to create such a record, in any event, whenever I observe
a significant *Code* violation. I then want to amend the directory at the
main distribution panel or service. Edison-base fuseholders get fustat
adapters, even if I haven't found overfusing. Prevention. Last, but not
least, I want to know why the fuse blew.

In at least some cases, I tell my customer that the subpanel needs
to be changed even when it is not illegal. If I actually work inside it,
this brings more responsibility. I won't add an additional circuit to
a loadcenter located in the back of a clothes closet—or an additional

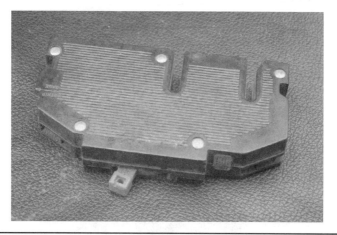

FIGURE 6-1 Zinscos had a bad reputation.

conductor to an existing circuit in that panel. I suppose that if the local inspector okayed the work, I might go ahead with it at the customer's insistence, but this would happen rarely.

Outdated Circuit Breakers

A far tougher call involves adding circuits to loadcenters that used to be legal, but are very dated. Those that come to mind include Pushmatics, Stabloks, and Zinscos. Figure 6-1 shows the general appearance of a Zinsco; Fig. 6-2—despite being embossed *ITE Sylvania* shows a Zinsco more from the rear, so you can see the busbar clip. Others are

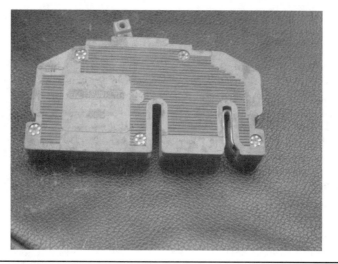

FIGURE 6-2 Zinscos were fed by bus stabs.

Figure 6-3 Pushmatics also had a bad reputation.

located in buildings that have changed so much as to stretch the concept of grandfathering to its limit. A (underlabeled) Pushmatic still in use at the end of 2009 is shown in Figs. 6-3 (outside) and 6-4 (inside). Thankfully, the oversized FPE bolt-down CB in Fig. 6-5 is one I haven't seen in the field in decades. One of the many reasons these can be challenging is that unreliable breakers may need to be left in place because the appropriate panel blanks are similarly exotic and hard to come by. Two varieties are shown in Fig. 6-6. They are just too different in size, Figure 6-6 shows CBs below the corresponding blanks. Blanks can't be universally interchangeable, as shown by Fig. 6-7's side-by-sides of the monster FPE bolt-down just shown, a Zinsco, an FPE Stablok skinny, a Murray (at least here the shape is standard), and a Pushmatic.

Classified Information
In dealing with outdated breakers and panels, sometimes I need to install another circuit breaker. This could be because one of the breakers

FIGURE 6-4
Pushmatics did
secure positively.

FIGURE 6-5 This design *looks* standard.

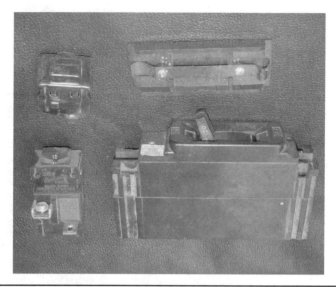

FIGURE **6-6** These took vastly different filler plates.

FIGURE **6-7** CB sizes and shapes differed greatly.

in the panel has gone bad or because the circuit it feeds is overfused—this could even be for a legitimate reason, such as a 20-amp circuit feeding K&T in a wall that was retrofitted with blown thermal insulation. It needs to be changed to 15 amps. It could be that there is an unclosed opening—even an unclosed opening resulting from an installer using a skinny rather than a full-width breaker. It could even be because I decide there is good reason to add a circuit.

If a brand or style has not been sold for some time, the odds are that OEM parts, such as even breakers, are out of stock. I consider using a Classified breaker to be by far the best choice. So long as a NRTL has Classified it for this use, it has passed the same test as breakers Listed for that panel—except that it may also meet modern standards, for instance, having a higher AIC: this is all to the good. One alternative is a takeout, whose calibration may have drifted. Another is a gray market breaker, meaning one whose history is very uncertain. Its calibration is unknown, and it may even be a counterfeit. Chinese counterfeits are a relatively recent development, but in the 1980s Mexican counterfeits were common. (For that matter, currently Chinese imports often enter the U.S. via Mexico.) Fakes are not a new phenomenon. "Reconditioned" breakers are something I stay away from. Their internal workings are unlikely to have been inspected or touched and besides, I know of no companies that remanufacture breakers so as to meet UL/ANSI standards and get their product reListed. Instead, my understanding is that they clean them up externally and perform very coarse tests to confirm that they will trip—not anything like confirming that their tripping curves match Listed specs.

Example   One bodega (small neighborhood grocery) that calls me in not only has a Pushmatic panel, a plain ITE panel of similar vintage, and a QO, but has a separate service to each one, because it combines three formerly independent stores! So far, I've lived with this situation as I am called in to repair their wiring. They understand the setup is bad; I halfway accept the arguments that this is not as hazardous as it might be. I informed the AHJ of the situation. He was willing to let it be, but he would not allow any of the panels to be replaced—not until the owners were ready to integrate them all into one service.

Inadequate clearance because of clutter is a common occurrence; but there are worse problems. Odd locations aside, I see many examples of even service equipment with inadequate clearance. In most cases, these passed inspection legitimately. Subsequently, though, the tenant or owner had equipment installed that encroaches on the required clearances. Storage basements are notorious for clutter, as are electrical closets that start out as dedicated spaces and end up doing double duty to hold cleaning supplies. Despite the fact that many plumbing/pipefitting firms carry electrical licenses as well, heating and plumbing installers seem to be among the most common violators. But electricians do it too.

The handy homeowner and the handyperson, of course, create their full share of the problems. However, regardless of who may have worked on the wiring last, I've learned not to make any firm assumption about how safe it will be.

Example   One nasty case of turning an installation from inconvenient to illegal and involving service equipment consisted of a fusebox, mounted low on the front wall of a house, in the kitchen. Despite its odd placement, perhaps 2 ft. above the finished floor (AFF), the fusebox started out quite legally. However, someone mounted a shelf above it and also extended a peninsula counter and cabinets out from the wall right next to it.

More and more, I've been learning to wait for clear working space, rather than work on equipment while impeded by clutter. Feeling unsafe, I won't proceed—even to inspect. Usually, the customer prefers to put it back the way it was; rarely will I myself restore illegalities that I've removed. As an inspector, I won't walk away from, or red-tag for, clutter—unless it prevents my access. I might set aside one or two items so I can get at a panel, but if it's in a closet as in Fig. 6-8, clear it!

Just before I finally went for my third IAEI certification, I heard Bill King, then representing the CPSC, and Phil Cox, then IAEI's CEO, talking about this at a board meeting of the old National Electrical Safety

Figure 6-8   Many utility closets are quite blocked.

FIGURE 6-9   No open KO is *really* ok.

Foundation. They agreed that an intrusion into dedicated space is an example of a battle they might choose not to fight. This pragmatism did not come as a shock. Figure 6-9 shows another one I'd let be, at least if the edges above and alongside the open KO were caulked. On the other hand, the multiple open KOs seen looking out from the panel in Fig. 6-10 were quite unacceptable.

FIGURE 6-10   With so many open KOs, I have to red tag.

When it has been necessary to correct an accessibility problem, one option I've used is conversion. When a subpanel hides in a ceiling or in the back of a closet, I've sometimes been able to fish multiple cables from it to a new subpanel at a more appropriate location. Then I blank off the old subpanel's cabinet, converting it into a large pull box. Doing so is cumbersome, but does not require structural changes. Note that some inspectors require adding screws to the hasp of a panelboard that has been converted to a junction box, to positively secure its door.

Example   When the loadcenter is illegal because access is blocked largely by a fixed part of the building structure, the solution is more difficult. One such installation involved a subpanel, in a different kitchen. I found two quibbling flaws and one real problem.

First, it was located not 2 ft., but a mere 2 in off the floor. Why was it mounted way down there? Probably to make it easy to fish cables up from the basement. There is nothing illegal about setting a load-center in that awkward location. (Yes, an inspector could characterize it as "not readily accessible," and red-tag it, but doing so would entail an idiosyncratic, nonstandard definition of "readily accessible," exceeding the intent of the NEC. In a flood zone, its location might have violated FEMA or similar rules. This was not a flood zone.)

Second, the cables, including NM, emerged from the floor without protection. To object to this would have been extreme. The cabinet installed 2 in up protected the cables fully from any source of mechanical injury that might reasonably be expected. In this setting, it blocked most access even though it was neither in front of nor surrounding them.

There was one clear illegality: a permanently installed bench, a sort of window seat, right in front of the loadcenter. Not only did the bench interfere with the clearance required under Section 110.26, it partly blocked the panel cover.

The best way to correct this situation would have been to relocate the window seat. Failing that, the fix described above—running cables to a new subpanel in, say, the basement, would have been a partial solution. A shortcoming of this solution is that NEC Section 314.29 requires access to wiring in junction boxes; this still would have been missing. Still, many inspectors would go along with such a partial solution. As it happened, the customer would not authorize any work to correct that violation. Therefore, I simply found the dead fuse and replaced it with a properly sized no-tamper fuse and fustat adapter.

I also marked the service equipment, to remind myself of the problem on the next visit, or to alert the next electrician.

Example   On another job, I found a service panel in a corner closet whose door blocked access to the left two mounting screws. Consequently, those screws had been left out. Two new holes had been drilled further in, for sheet-metal screws to hold the cover. In this case, I told the customer that the closet door and door frame had to be removed, so that the panelboard could be restored to the manufacturer's design. Perhaps a clever carpenter could design a pocket door arrangement that would avoid blocking full access to the panel and its cover. (I have no idea whether the customer chose to have this done.)

Example Especially in nonresidential installations, it is critical to maintain the integrity of panel cabinets as investigated. This does not rule out sensible accommodation. In one commercial panel, one of the original cover screws had been replaced by one of the very same style, from the same manufacturer, but longer, shown in Fig. 6-11. I had no problem with this.

Example Inspector and instructor C. L. Griffith described a case he observed at an institutional job where he noted inadequate internal equipment clearances. Switchgear busbars had suffered arcing damage as a result. The problems revolved around insulating boards that supported the physical clearances, ensuring electrical separation. Some of the boards quite evidently were missing, based on the bolts that should have supported them. Others had lost any insulating properties, becoming somewhat conductive, and contributing to the arcing. The ventilating system that removed heat generated in the switchgear had drawn water into the gear, leaving those insulating boards saturated with contaminants.

FIGURE 6-11 I'll accept an OEM screw a tad longer.

Complex Kludges I've Encountered

Multiwire Mischief

The next odd situation found in older buildings is not nearly as unusual as I wish. There are quite a few rules about multiwire circuits. Every one of them has been trampled upon. Multiwire circuits, by rights, consist of two or (in three-phase systems) three energized conductors sharing one neutral.

The neutral should be identified. Energized and neutral wires should run in the same cable or conduit. The neutral should be more lightly loaded, most of the time, than it would be in two-wire 120-volt circuit serving the same loads—and never more heavily loaded. Under recent editions of the Code, all discontinuities in the neutral should be connected in splices, pigtailing it at daisy-chained outlets, rather than feeding an upstream neutral into device terminals, and a downstream neutral out of them. The energized conductors should originate from different busbars in the same panelboard. The overcurrent devices should be arranged so as to be disconnected simultaneously, if they feed line-to-line loads. Whatever their use, they must be equipped with means for simultaneous disconnection. By rights, this is how it should be done these days.

There are three differences between that modern design and the multiwire setups I tend to find in older buildings. First, Section 210.4 has only required simultaneous disconnection universally since 2005. I know that a circuit I've turned off may be backfed through a shared neutral—either legally or ignorantly or inadvertently. For a preliminary (but only preliminary) indication that I may be safe from this risk, I look inside the panel. If the conductors of the circuit I'm working on leave as two-wire circuits, they won't pose this danger—*in the absence of violations*. This is also true if they leave as multiwire circuits and I've ensured that all the energized conductors feeding the circuit from this panel are deenergized, ideally locked out and tagged out.

Second, daisy-chaining a multiwire circuit's neutral through device terminals was legal until adoption of the 1971 NEC. If for no other reason than that, in energized testing I'm conscious that one loose connection in a 120 VAC circuit could result in 208 or 240 V being divided by whatever loads are connected; these loads could include fingers, in a careless moment, though the voltage to ground remains 120.

Third, as with so many aspects of older buildings, the longer a system has been around, the more likely it is that something went bad and was kludged back into operation. In the case I describe in the following—and this can be a very complex type to troubleshoot—some wire or connection failed and the replacement was "borrowed" from a convenient nearby outlet.

Some hazards I've run across had an even simpler origin: someone tied one circuit's return conductor with that of another or others somewhere downstream—"all whites go together." They created a multiwire circuit, possibly with a paralleled 12 or 14 AWG neutral.

In old installations, I've often found cross-connections between neutrals and grounding conductors after the main bonding jumper—sometimes, perhaps even more often resulting from confused use of neutral bars at subpanels. Bootleg grounds, if not also other downstream grounding, means that any voltage imposed on the neutral of a multiwire circuit might be present—perhaps hazardously—on the exposed surfaces of electrical equipment.

Example   An example or two will demonstrate that problem of "borrowed" lines. They are quite complex, but they do represent real installations. I invite you to sketch them out for yourself, putting yourself in the place of the person troubleshooting.

A large two-gang switch box—call it $\underline{A}$—controls two lighting outlets, and is fed from one of them. $\underline{A}$ contains three two-wire cables, $\underline{a}$, $\underline{b}$ and $\underline{c}$, and one three-wire cable, $\underline{d}$. The three-wire cable, $\underline{d}$, contains, first, a switched conductor going to one of the lampholders or ballasts; second, a neutral connected to a splice that feeds both lampholders or ballasts, one via $\underline{d}$ and one via $\underline{a}$, one of the two-wire cables; and, third, an energized wire that is spliced to the black wire in cable $\underline{c}$ and pigtails to the switches. The white wires in $\underline{b}$ and $\underline{c}$ are spliced together. The black wire in $\underline{a}$ is the switch leg to the second light. The black wire in $\underline{b}$ is disconnected. In the electrical panel, the cables come from circuits 23 and 35. Originally, $\underline{c}$ and $\underline{b}$ had been connected to circuit 23, and $\underline{a}$ and $\underline{d}$ to circuit 35.

Apparently, someone was called in because power to part of circuit 23 had been lost. Power still was getting to the light. Energized conductor and neutral both still came through on circuit 35 to point $\underline{A}$. Power continued along circuit 23 to point $\underline{B}$, the next outlet upstream on that circuit, but did not get from there to the switch box, A. However, Mister fix-it couldn't—or at least didn't—put in the time and effort to find the fault upstream location of the fault.

Here is what my "get it working" predecessor did, from what I could tell:

> disconnected the black wires in the unsatisfactory two-wire cables, $\underline{b}$ and $\underline{c}$, at point $\underline{A}$.
> determined which one continued on to $\underline{C}$, the dead outlet downstream. This was $\underline{c}$.
> capped the upstream black wire, $\underline{b'}$, from circuit 23 at $\underline{A}$
> brought $\underline{c'}$, the black wire heading downstream, into the splice with circuit 35's energized wires.

The result was that the one energized wire from circuit 35, $\underline{d'}$, now fed into two, parallel, neutral returns. While circuit 35 is energized and serving loads, workers have some chance of receiving a shock should they interrupt either neutral and get in series with the resultant upstream and downstream segments. Because they would be in parallel with the other neutral, that risk is not as severe as sometimes. At least equally troubling is the fact that the conductors now carrying the power of this circuit are following different paths. Inductive effects will be particularly severe where metal raceways or enclosures are used. If both energized and neutral conductors of the two-wire cable going on downstream to point "C" had been disconnected from the dead upstream two-wire circuit 23 cable and changed to the circuit 35 splices, this would not be a problem, though there still would be a violation if the circuit directory was not changed with respect to circuit 23 and circuit 35. Given the puny space that most circuit directories provide, even well-intentioned efforts to amend them have difficulty staying legible. There could still be two potential problems. First, is it permissible to add receptacles to circuit 35? It is illegal, for example, if it serves lights in multiple bathrooms. Second, does point C serve a heavy load? I have been to jobs where that sort of rearrangement had overloaded a circuit. This means that now the circuit carried power—but it kept tripping.

The problem created by borrowing power is less serious than that created where the circuit 23 neutral had the break, and so the circuit 35 return path is the one that's borrowed. I've seen several such cases where an ignorant installer brought the neutral heading downstream into the splice containing the other circuit's neutrals. Doing so turned the home run into a multiwire circuit—at best. Legalities aside, the one neutral from circuit 35 now was fed by two, parallel, energized conductors. Workers have a chance of being shocked by interrupting and getting in series with the neutral when either circuit 23 or circuit 35 is energized and serving loads. I chose circuit numbers indicating that given their spacing, it would have been physically impossible to use handle ties, not that this always was required the way it is today. Moreover, someone in a hurry would be even less likely to recognize the need for simultaneous disconnection before work on either circuit. Even worse, depending on which actual circuit numbers are involved, sometimes these kludges originate from the same busbar, double-loading the return conductor.

The identical type of problem has been created when, instead of stealing power from another circuit passing through the same box, someone borrowed power by running a cable over to a nearby outlet that was on a different circuit. In other words, if box A did not contain both circuits 23 and 35, they might have run a cable over from box A, which contained circuit 35, to box C, which had no energized conductor, having been fed by circuit 23, the circuit that was interrupted at some location that remains unknown.

One difference is that when running a cable from a different box, two unbroken conductors are available, so there is some intuitive inclination to run both energized conductors and neutrals, rather than continuing to use one of the conductors from the defective circuit. I don't know why this doesn't happen more often.

The other case where such "borrowing" occurs without creating a hazard is where the two circuits sharing an enclosure are, in fact, a legitimate multiwire circuit. Instead of being controlled by breakers 23 and 35, they are far more likely to be on breakers 21 and 23; they may even be handle-tied. This certainly is not a given. This set of scenarios underscores two points. First, the previous repair person may have made an assumption that created a hazard, without the slightest awareness. Second, when I don't watch out for the hazards that have been created by such mistakes, I endanger myself.

I've seen similar, but rarer, cases of modifying a multiwire circuit and overloading the neutral, where a fuseholder, let's say fuseholder 9, became unreliable and the conductor attached to it was moved to a free fuseholder, fuseholder 11. In other instances, where all fuseholders were already occupied, the "orphaned" conductor was spliced to the wire that had been attached to a good fuseholder, and the splice was pigtailed to the good fuseholder. Other times, the conductor was moved to lie, illegally, under the same fuseholder's screw terminal as another wire. When the wire that was moved had been part of a multiwire circuit, I've found the neutral double-loaded, potentially carrying 30 amps to the energized conductors' 15.

When circuits are improperly mixed, it is far more likely that the result will induce current in conduit and cable armor.

Polarity Gone Amuck

Consultant Earl Roberts told of a small New England town, built before equipment-grounding conductors were required, whose AHJ instructed installers to use white as the energized conductor and black as the neutral return. Unusual.

When I have occasion to work on wiring with reversed polarity identification, I am very careful indeed. Who's to say whether the energized or the return conductor got switched, heading to lighting outlets, or whatever was done with respect to multiwire circuits. A similar, but less severe, problem arises when conductors are old enough to lack color identification.

In all these cases, I do try to determine which conductor is energized and which is the neutral. When I am not feeling too pressed for time, I wrap black or white tape around each existing conductor, as appropriate. I remain uneasy about the confusion I may create for future workers if other parts of the circuit are clearly identified the other way around. Despite this, where my tester tells me which conductor is which, I consider bringing order to be a plus.

Downstream Grounds: Returns to Haunt

The ignorant mistakes I've described so far, including the multiwire miswiring, account for only a few of the circuits whose current returns along unexpected paths.

Some horror stories stem from raceways deteriorating to the point that the systems feeding them may be grounded, but parts that once were grounded—perhaps were grounded from day one—are no longer reliably grounded. For instance, I understand that after being installed in a concrete slab treated with a common additive, calcium chloride, after 30 years even heavy-wall, rigid metal conduit has completely corroded through! Also deadly. Figure 6-12 shows unprotected electrical metallic tubing (EMT) rotted away by the slab from which it emerges. Perhaps the corrosion also caused the shifting that pulled it out of the box connector, but I suspect that it is more likely not the cause.

FIGURE 6-12   Concrete can eat EMT.

Example   Donald Barrett, former Chief Electrical Inspector for Prince George's County, Maryland, told of a death caused by someone touching a defective air conditioning compressor while standing outdoors on the slab through which the compressor's circuit ran. The slab was grounded not only by being on the ground, but also by the conduit running through it. Ironically and fatally, the makeup of that slab had destroyed the section of conduit it enclosed, so that the circuit's grounding was interrupted before the conduit reached the defective air conditioner. Enough fault current flowed through that slab and the ground to kill a person, but not to trip the breaker.

A three-prong tool with the third prong removed, or with a cheater employed on the cord, can be far more dangerous than an old-style two-prong tool. For the same reason, three-prong equipment connected to a circuit that has lost its ground presents increased hazard, because all accessible metal parts not intended to carry current are bonded together. The result is that any ground fault energizes all these parts. If the upstream grounding on the circuit, rather than just at the one tool or outlet, is defeated, one piece of shorted equipment will create a shock hazard at all three-prong equipment on those parts of that circuit that remain bonded together—even undamaged equipment connected to other outlets!

In the previous example, the concrete that would be expected to protect the raceway destroyed it. While third prongs are intended to make equipment safer, when the system is defective they too can have the opposite effect.

Certain other changes in how electricians wire equipment will not create any surprise for readers of this edition, but are worth mentioning for the sake of future readers.

Ranges and Dryers

The neutral and grounding conductors generally are to be joined at one place only: at the main bonding jumper, or its equivalent at a separately derived system or, frequently, at a separate building. The neutral is expected to carry current, but the grounding conductor never is intended to be energized except, momentarily, during a fault. However, the exceptions in NEC Section 250.140 used to be broader. This means that in older homes a fair amount of current may travel over the exempted (concentric) grounding conductors.

NEMA 30R and NEMA 50R receptacles, intended for, respectively, electric dryers and electric ranges, offered combined 250/125-Volt outlets, but no separate ground. The chassis of any appliance using a corresponding NEMA 30P or NEMA 50P cord connector was bonded to the third prong. Type SEC service cable with a concentric neutral, covered but not insulated, was permitted as the feed, provided that it originated from a loadcenter that was also the service equipment. These were some of the *rationales*:

- that the neutral would tend to be large enough to suffer sufficiently low-voltage drop that it would remain safely close to ground potential;
- that it would be less likely to break contact than a grounding conductor connected only to the frame; and
- that at least if it broke, operation would be affected, so repair would come more quickly than with a grounding conductor that might lose continuity without this being unnoticed.

This design was allowed by Special Permission originally, and only for ranges, and then graduated to a straightforward rule in the 1942 amendments to the 1940 NEC. Dryers were added in the 1953 NEC. This is how it remained for decades; the general permission was not removed until the 1993 NEC. The *purpose*, I've been told, was to reduce copper use and thus save metal (an alternate view is that the purpose of the exception was to make electric dryer installations more competitive with gas dryers.)

Plug-in appliances were not the only ones wired with the ground and neutral combined. In early 2010, I was called in to remove a 1962 installation of stove and cooktop, each fed by 10-3 conductors, and grounded to their neutrals. They left the panel (Fig. 6-15) in BX and transitioned in junction boxes to 10-3 with ground (WG) NM as shown in Fig. 6-13. They bore factory tags (shown in Fig. 6-14), and had been

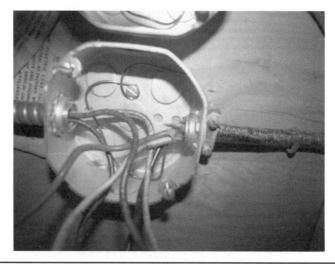

FIGURE 6-13 Out-of-the-way boxes often show bad bonding.

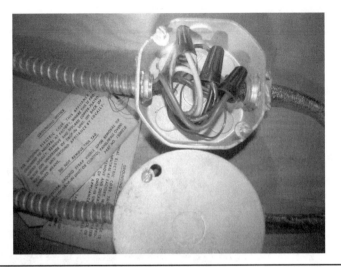

FIGURE 6-14 Cooking units fed by 10 AWG conductors could use neutral grounds.

installed *almost* legally. Two of the problems are that they were bonded to mounting screws (Fig. 6-13), and that their covers overlapped so that one cover was held slightly away from its box (Fig. 6-16).

I frequently find a subpanel that was formerly the service panel still feeding a NEMA 30R or NEMA 50R receptacle via SEC. Because of the panel's demotion, the same SEC run is now illegal. A metallic

FIGURE 6-15 A 240-volt oven and range on plug fuses was illegal in 1962.

Figure **6-16**
Overlap meant one
cover didn't close
fully.

Figure 6-16
Overlap meant one cover didn't close fully.

junction box that's bonded to SEC on a branch circuit certainly is illegal.

Antiques

I've seen some additional types of equipment that ground to the neutral or, worse, that substitute the grounding means, whether that ground is a raceway or a conductor, for a neutral. Some were easy to upgrade, while others were a trial.

Merely Odd and Antique

Some equipment is okay and can be left working. It can be replaced when it dies. Other items, such as the old switch in Fig. 6-17, probably will need to have their functions relocated, at least when they stop working. A "120/240—your choice" receptacle, seen in Fig. 6-18, was ungrounded and should be replaced by a receptacle suitable for its purpose and location when it dies.

Crowfoots

First among the bad ones is the crowfoot. It's a standard-sized receptacle, single or duplex, whose openings look almost like and whose energized slots are configured exactly like those of a NEMA 50R-range receptacle, except for being smaller. Like a NEMA 50R, the crowfoot

FIGURE **6-17** To replace the switch, I must replace the box.

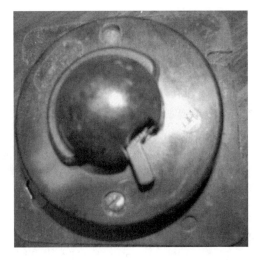

FIGURE **6-18** This fed 120 or 240 volt, not both together.

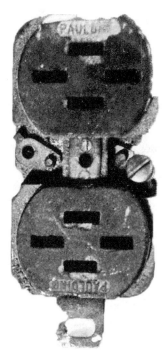

has no separate ground. I've only once found a crowfoot receptacle in use—nothing made in recent decades will plug into it. That particular outlet fed a standard two-prong plug whose prongs had been twisted brutally to make them go in.

When the wires feeding a crowfoot were in good condition, I've substituted a modern receptacle. If the enclosure was grounded, that did the trick. If it was not grounded, replacing it was like the replacement of any ungrounded receptacle, with one difference. If it happened to be a dedicated outlet, I had a spare wire. I have been permitted to use it as a grounding conductor, reidentified where accessible. Technically, in the applicable locations, this gambit did require formal Special Permission, in the case of 12 and 14 AWG conductors.

Sheer Ignorance

Crowfoots constitute the more promising case. Other arrangements, where someone used the ground as a return simply because doing so made the outlet work, can be beyond any reasonable repair.

Bootlegging the ground as a return often creates hazardous conditions. In these cases, the arguments that wangled an exception for ranges and dryers did not apply. In some of the worst, the danger is compounded because not only was the electrical ground used as the return, but the ground used to rely upon the plumbing system for continuity—generally legal until adoption of the 1993 *Code*, when Section 250-50(b) was updated. When someone replaced a damaged metal water pipe with plastic, the outlet went dead. If the light or appliance had been removed or at least turned off, fine. Otherwise, if the plumber didn't bond around the section as he worked, he was at risk if a load was connected. If not, later users were at risk.

Three More Hazards

I've found many other wiring peculiarities in old buildings that are not necessarily a problem; some, at least, are not technically violations. All could have proved hazardous to me before I understood what I was dealing with.

Buried Splices

Knob and Tube (Initially Discussed in Chapter 3)

I have encountered many instances of amateurishly and illegally created buried splices. Knob-and-tube wiring like that in Fig. 6-19, of course, had legally buried splices. However, I haven't discovered any professional K&T splices that went bad. Dealing with K&T, I wouldn't worry if I couldn't find a splice. There is the question of what to do when work is needed on existing K&T installations. Since extension

FIGURE 6-19 Loom could protect wires from metal edges.

of existing knob-and-tube circuits is authorized explicitly by Section 394.10, repair certainly is legal.

Before choosing to do so, I have to answer two questions. First, do I want to be repairing or extending a circuit that, in most localities, was installed without a ground, and furthermore an installation that very possibly is falling apart after being in place for so many years? If the conductors have been covered with thermal insulation—unlike the interior-wall switch leg in Fig. 6-20—they could very easily have suffered from overheating, on top of simply aging. True, without over-heating, rag wiring in open installation can survive fine when properly protected by tubes. (Fig. 6-21; Fig. 6-22 makes it clear that tubes are basically woven cloth but they're tough enough for the job.) Second, how much trouble will it be to get hold of the necessary parts locally; or is it worth having them shipped? They are available, though not as widely as the very old ad in Fig. 6-23 suggests was once the case. Using individual knobs isn't necessary, either, as shown in the layout in Fig. 6-24. However, a knob can lift and redirect each wire to its tube, as in Fig. 6-25. The tubes separate wires from either flammable or, as in Fig. 6-26, grounded items.

If a customer decided that a break in an ancient circuit run in knob-and-tube or open wiring was to be repaired, and I judged that its insulation was healthy enough to permit that repair, I'd bring the knob-and-tube wiring into a nonmetallic box and splice a cable onto it in that box. From there I'd run cable back to the panel containing the overcurrent device, rather than merely adding a grounding conductor.

At one point, K&T and open wiring seem to have been a default wiring system, as witness the installation guidelines from a wiring book (Fig. 6-27). That's long ago—1917. I understand there is one ex-ception to the idea that concealed knob-and-tube is a thing of the past.

FIGURE 6-20 K&T switch wires could come from two directions.

In areas subject to severe flooding, I have been told, concealed knob-and-tube wiring is common, because it retains an important advantage: it dries out faster than does cable or raceway wiring. I have never talked with someone who'd done this work themselves, so I don't know anything more; and it's not practiced in jurisdictions where I'm licensed.

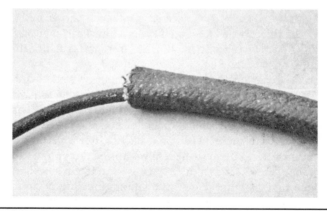

FIGURE 6-21   Loom offered thick protection.

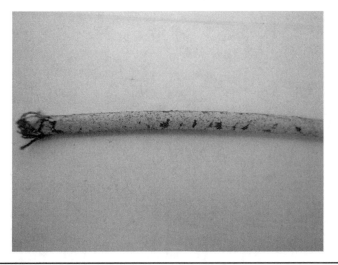

FIGURE 6-22 Loom was at core a tough, fabric tube.

THIS IS THE NEW

"**Buckeye**"

"*Trade Mark*"

Split-Knob
Insulator

Officially Approved for use in Knob and Tubework
Described on Page 131 in this issue of
"STANDARD WIRING."

It is the only knob with two available wire grooves and
the interlocking feature, which keeps the pieces in place
while being installed. Has liberal screw protection and
may be installed with screws, or nails, where approved by
inspectors. Note the triangular construction of the wire
way, which grips the wire absolutely, without injuring
insulation.

This knob is strong and substantial; neat and compact
in appearance; and complies with the Underwriters' rules.

Samples, Printed Matter and Discounts on request.

FIGURE 6-23 Knobs came in various designs.

FIGURE 6-24   Separate knobs aren't required.

One last word about knob-and-tube wiring. It was intended and permitted for use in free air, not buried in insulation. That's why 12 AWG conductors in K&T, even grandfathered ones, need to be derated to 15 amp.

Illegally Buried Splices

Other buried splices are just violations, either created by someone doing electrical work or by someone who came along later, such as a plasterer. To locate such splices, I look for new walls, added by those

FIGURE 6-25   Knobs enable wires to change direction.

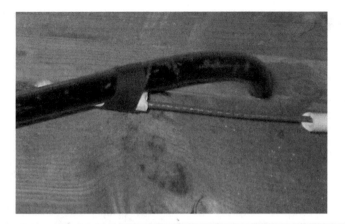

FIGURE 6-26 This looks unorthodox—but may not be required.

ignorant of the problem. I look for dimples in ceilings (Fig. 4-10), roughly the size and shape of a 1900 or an octagon box (8B) or, in walls, of a switch box. I haven't personally had success at using a voltage tracer for this. On one job, Bruce Moore, a sharp field engineer for Met Laboratories, found a buried connection by using thermography.

In the worst case, where I just haven't found the missing connection and it's not been worth devoting more time, I've disconnected the line-side wires, at least one of which had been interrupted somewhere unknown, capped them, and run fresh cable. If I could figure out from the panelboard directory which loads were involved, or identify additional nonpowered loads, I might also have tried to disconnect and

> Solid porcelain knobs should be used at the end of runs where circuits are terminated. Split knobs or cleats should be used for conductors smaller than No. 8 B. & S. gage, except at the end of runs.
>
> All knobs or cleats should be fastened by screws or nails of generous length. If nails are used they should be long enough to penetrate the woodwork not less than one-half the length of the knob and fully the thickness of the cleat. Washers should be used with both screws and nails to prevent injury to the knobs or cleats.
>
> **Splicing** should be done so as to make the wires mechanically and electrically secure without solder; then they should be soldered to insure preservation
>
> 78

FIGURE 6-27 By 1917, K&T installation was standardized.

Figure 6-28   Tidy or not, this boxless transition was illegal and unsafe.

cap the wires heading "nowhere" at the last accessible box to the other side of the hidden interruption. However, exhaustive labeling is very much the exception. Continuity in the neutral often can enable me to trace the circuit back, at least when neutrals have not been confused by previous workers.

I have two reasons for disconnecting and capping. First, the customer might subsequently discover another outlet that was mysteriously dead. If it turned out to be the one to which the capped wires went, I can splice them in to the new cable and restore more of the circuit. Sometimes, this will even turn out to have a bad connection which I can repair, letting me restore the circuit to its original configuration. Second, I want to interrupt anything that feeds or backfeeds into a bad splice. Even if it wasn't hazardous, it could be responsible for an intermittent.

Another problem that I've learned to associate with missing connections is loss of grounding. Especially, but not only where the work has included running splices, such as the transition of BX into NM that I show in Fig. 6-28, the grounding continuity of the armor is lost. Ignorant people attached the BX's aluminum bonding strip to the NM's grounding wire. The only up side to their doing so is that if there is a significant fault, if their splice is good enough the flimsy strip probably will vaporize too fast to ignite anything, in addition to not causing the overcurrent device to operate.

Unexpected Power

I've also found power when nothing should have remained energized. The cause could be more outrageous than any I've run into so far, and I've seen some bad ones. I've heard of too many cases of theft of utility power, sometimes bypassing normal OC devices. Normally, if I want to work in a residential or small commercial service panel unenergized, I pull the meter. Even so, I always test after pulling the meter.

Ignorant installation and use of backup generators is a related issue. I understand that one common result is generator backfeeds.

Example   A particularly unfortunate example of insufficient customer education may have come about because equipment sellers had no idea of the NEC rules or the Listing restrictions. The customer had an increasing need for reliable power, because of his worsening physical condition. He had been sold a backup generator that was unsuitable for his situation and was installed quite illegally. However, I am sure that the installer was able to demonstrate that it worked. Unfortunately, what apparently never was mentioned is that the unit needed to be exercised weekly. As a result, the customer had left it sitting dormant until he had a power outage. Then it didn't work and he called me in.

Returning to unexpected power, I have found this all too often when the lines in separate tenancies are not properly separated. There is more than one service where originally there was only a single service. In providing each occupancy with its own meter, MAIN, and distribution panel, the existing circuitry was not segregated as well as it should have been. I'm not thinking of large industrial establishments where there are multiple sources of power, because that's not where I have experience.

I see this in apartment buildings whose tenants have separate meters and apartments have loads that are not controlled by the load-centers dedicated to the apartments. Sometimes this is because common building circuits entered their premises, circuits intended, say, for alarm systems or even for hall lights. On other occasions, power from Apartment 24 serves loads in Apartment 23. Sam Levinrad, a vastly experienced consultant and retired inspector, also has found this situation to be quite common. I've found the same thing at small commercial customers' premises, and in at least one institution. At my insistence, the bodega that had combined three prior tenant's buildings put signs at each of the three service panels alerting users and first responders to the presence of the others—served by separate meters (shown in next chapter's Fig. 7-10).

My Response to Finding Power Present When It Should Be Absent
Whatever the reason that unexpected energized conductors are found, the fact that they do turn up underscores one rule that is even more important to heed in work on old buildings than in new: I know that power is off only when and where my tester tells me power is off and I confirm that the tool's working.

> When I was an apprentice, I was proud of my "wire-stripping," meaning notched-blade, lineman's pliers. More sober these days, I am proud when none of the pliers I own tells the story of having accidentally cut energized cable.

To be realistic, I have to offer two levels of solution for intermingled power. In the case of apparent theft of power, I'm out of there. My

Figure 6-29
Ty-rapping to
plumbing is not
legal support.

inclination is to notify the serving utility, if the present occupant won't
do so after I warn them. More generally, the least I do when I've told the
customer and they don't authorize me to fix problems that constitute
potential hazards, is to mark the service equipment with a permanent
warning.

With the go-ahead for a real fix, I start by removing jumpers or
bypasses and closing any openings this leaves. I want to close existing openings that should be closed generally. This is normally higher
on my list of priorities than adding missing cable supports and replacing misapplied supports like the doubly wrong one in Fig. 6-29
(between the camera and the open KO). I want to disconnect power
that improperly originates in a different occupancy. I find it better to
blank off any outlets served by the wrong meter, rather than simply
refeeding them from the proper loadcenter and capping the foreign
wires, even if there is sufficient box volume for that. I don't like leaving energized wires in an occupancy that don't belong there, at least
not in any box someone is likely to have reason to open.

Unfortunately, work authorization in many of these occupancies
is in the hands of people who focus on whether something's working
much more than on *Code* compliance or on "abstract" danger—until
there's a tragedy or near-tragedy.

Example: Mystery Voltage   To close this section, here is an one odd little story about mingled electrical systems. I was called in to troubleshoot the system at a townhouse whose owner had gotten a tingle in the shower. No voltage registered on my Wiggy^R. I switched to a wider-range multimeter and found 10–13 VAC between various points—the shower head, the faucet valve, and the metal drain grille. At first it seemed that the plumbing, perhaps the cold water line, was energized. Not so. Nor was it the shower rod, which was energized, I understand, by NM damage in some U.S. military quarters in Iraq. Further testing, using a properly powered three-prong extension cord for reference, confirmed that a potential difference was present between the drain grille and ground.

After pulling the meter and confirming in the presence of a utility representative that no power was coming through the house's service, I continued to measure voltage between the drain and the extension cord's ground. The drainpipe itself was plastic, but power was transmitted through the sludge and slime coating its inside—even with the main turned off at the service. The origin was outside the house: the neighbor's house was sending some voltage to ground through the plumbing system rather than through their incoming service neutral.

> Throwing my amprobe around the GEC and finding nontrivial current flowing can warn of a poor service neutral connection IFF the main bonding jumper is effective.

To my mind, this story illustrates how sometimes, in order to track a problem to its source, I've had to go beyond the bounds of the tenancy or even the building. In the case of this townhouse, when the trace voltage went away shortly afterward, the owner let it be. Otherwise, we would have insisted that the utility do something about it. Bringing in the power company's representative, as well as the neighbor, would have been an example of notifying those concerned. It would have formed the basis for a subsequent demand that they take steps to eliminate the stray voltage.

Example   That story, like the next one, also illustrates how I don't dare hold to the usual idea of a source of dangerous voltage, not when thinking about safety. Inspector Jim Wooten tells of a construction site on which he found a pile of dry plywood that registered 120 VAC to ground. The rosin in it was conductive enough to carry the power through the plywood from the metal siding it lay against, siding that had been energized by a nail penetrating NM. It wasn't a good enough circuit to trip the breaker, but plenty to shock a worker.

Unlisted Equipment

Another problem that is not unique to old buildings, but that I've found more commonly in older structures, is the use of unListed equipment. Historic buildings may include fixtures built before widespread Listing. Especially, but not only, where people or businesses occupy an old building because that's all they can afford, too often I find that over the years lots of corners have been cut. That has often had worse consequences than holding onto ancient equipment. Nonetheless, limping along with worn-out gear is dangerous.

I don't think this applies to all very old equipment, but I'm not trained to be a NRTL field engineer. That's what equipment evaluation really would require. All I've been able to do when faced with old, unListed fixtures, and ones that simply lost any labels, is use my seat-of-the-pants judgment, case by case, but give lots of warnings about the limits and uncertainty of any decision. NFPA is likely to adopt standards 790 and 791, addressing product field evaluation, in 2011.

Indoor Meters

The last old design I'll talk about is inside metering, as in the minimal service in Fig. 6-30. This example was still in good shape, as shown in

Figure 6-30 This house was served by 30-amp, 125-volt, one-pole.

Figure 6-31 Two 15-amp circuits were enough.

Figs. 6-31 and 6-32. In some cases, inside meters look just as safe and professional and guarded as anything modern. However, even there they may make it more difficult to provide the meter reader with ready access. Worse, there is more unprotected power inside the building before the first disconnect. It is also not as simple for firefighters to disconnect power before entering the building.

Certain cases are much worse. I've seen open wiring, minimally protected by loom, dive through wooden boards on which the meters are mounted. From behind the boards, the wires went into a trough or wireway. The problem with these setups and some other types of old inside metering is that they offer no real mechanical protection for the wires as they leave the meters.

I've found this in very old multiple-occupancy buildings. Each apartment might only have 30-amp service. With gas cooking and commonly supplied climate conditioning and hot water, at one time none or few of the tenants needed additional power. That's changed, but their services haven't.

In a building like this, I'll gladly retrofit GFCIs, if the wiring is safe enough to work with. I'll be happy to install fustat adapters. But adding loads is a problem. Even so, I often will add receptacles. The reason is simple. Ethically, I think it's better to give a tenant a

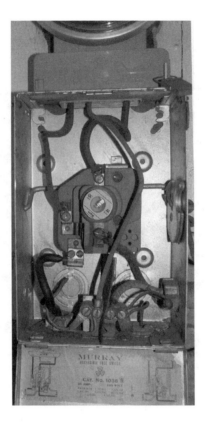

real receptacle for their new load than to leave them using extension cords.

It's still hard to convince anyone to authorize the major effort that would be required to upgrade the building's service. When it is authorized, I've seen it done pretty badly, on the cheapest of the cheap. On the other hand, the branch circuit wiring usually poses a worse hazard to the occupants. Upgrading this is a far bigger and messier job. Again, it tends to be done on the cheap. I've been called in repeatedly for troubleshooting and repair, where this had been done piecemeal and amateurishly (and without adequate power to feed the new loads).

Aluminum

The next big issue is aluminum wiring. Non-listed for interior wiring for more than 60 years, it was finally listed in the mid-1960s, when copper shortages drove up the price of copper conductors and aluminum saw real use on branch circuits.

Viewpoint Story   I was at a swimming pool, with my gear in a conference giveaway that touted copper wiring. Someone asked me, "What else is there besides copper?" I explained that electricians use aluminum and that, while it's less expensive, it is also less forgiving. I understand that a few jurisdictions prohibit any use of aluminum conductors in interior wiring. In the jurisdictions where I'm licensed, installation of aluminum wiring is restricted in residences and in smaller gauges.

I'm more nervous about aluminum than copper in branch circuits. It was easier to make connections wrong—if the worker didn't know what he was doing. I've seen this repeatedly. When copper arcs to copper, the terminations frequently fuse and melt together, forming a bolted short. When aluminum arcs to aluminum, I understand, it may well create a somewhat high-resistance connection to ground or to any other conductor in the vicinity. This problem is in addition to aluminum's suffering more temperature creep than copper and low conductivity after it's been exposed to air—as occurs at a conductor's surface in a very loose connection.

Heavier Cables

There are other issues, problems peculiar to older installations using larger aluminum conductors. Until the 1970s, electricians were not aware that lugs designed for copper conductors are not very compatible with aluminum and aluminum alloy conductors. With these installations, I want to replace the lugs. Sometimes, though, the equipment, as opposed to the lug, is marked explicitly for copper only. I haven't yet needed to add copper pigtails, except in branch circuit wiring. I don't even own a Hypress^{TM}.

If I'm working on the cables or terminations, I'm responsible to correct violations. Figure 6-33 is a violation I would correct in normal circumstances, using a Listed splicing device and means of bonding,

Figure 6-33   Multistranded (is not the same as) multiple wires.

and perhaps a terminal bar with sufficient openings for the small conductors. However, Fig. 6-34 not only shows two grounding conductors in an opening meant for one, but the smaller one is copper and the stranded, aluminum; I absolutely have to correct it. I won't retorque connections that merely are suspect, although I will make sure they are not actually loose. Retorquing can cause conductor creep, a concern I discuss further in the next chapter. If a connection has been running hot, I want to come in with a new lug on a newly stripped conductor end.

Ravindra H. Ganatra, a senior Alcan engineer, was the first to tell me there is no need for antioxidants with aluminum alloy. Alcan representative Tom Edwards reported that they recommend wirebrushing, adding antioxidant, and torquing when making up aluminum cable connections. However, this is not part of the Listing requirements. Simply making the connection up tight is acceptable. According to Dr. Jesse Aronstein, P.E., a materials scientist whose electrical studies are extensive, the conductors should be scraped and immediately coated before splicing or terminating. This matches what I've been taught by installers I respect. Still, I wouldn't require it as an inspector.

Aluminum Branch Circuit Wiring

As for aluminum branch circuit wiring, thousands of miles of the stuff remain in use, including some in my area. As is commonly the case, most of it was installed in the 1960s and 1970s. The original Grade 1350, installed up to around 1973, was worst: if the conductor was nicked, it was relatively likely to break.

It's Not Al All the Time

I do frequently need to clarify one thing for customers: not every shiny silvery wire is aluminum. Plenty of BX was manufactured with shiny, silvery tinned copper wires, the solder coating added as a barrier because copper and rubber interact. To check, I just scrape the tinned wires and the copper shows. Another clue is that tinned copper is stiffer than any aluminum of the same gauge.

And It's Not Always Copper!

Several styles of aluminum wire were installed and they didn't all look the same:

- early, lightly alloyed brittle wire;
- later alloy aluminum;
- copper-clad, a fairly rare compromise between the better price of aluminum and the better electrical characteristics of copper;
- nickel-clad and tin-clad, much rarer versions.

While it may technically violate 110.3(B), generally accepted practice allows use of any old twist-on connector with copper-clad, since what makes contact in splicing is the copper. I wouldn't be inclined to do so, but I also wouldn't criticize an installer I was inspecting for relying on them. If I looked at a cut end and saw aluminum, I'd be careful its ampacity limits had been respected and also be mindful of its greater fragility.

Repairs

My Main Concern   To begin with, if my advice is asked, I don't automatically say that existing aluminum branch circuit wiring requires repair or replacement. This said, even if there had been no problems associated with a building's aluminum branch circuit wiring, and no violations, I would strongly recommend adding combination-type AFCIs.

There are issues that I have not yet broached. When aluminum branch circuit wiring began to replace copper, the copper price increase meant the price of brass increased, so steel took the place of brass in the screws at device terminals. The chemical interaction between steel and aluminum is worse.

I'd advise replacing any original devices with CoAlrs (or Al/Cus), period, if all the aluminum wasn't suitably pigtailed with copper. Usually, beyond just swapping in new devices, there's also a need for some splicing of grounding or neutral conductors. This is where mistakes are often made.

Making Connections Involving Aluminum Branch Circuit Wiring

Copalum   There is one unquestioned reliable way to attach other wires to 12 and 10 aluminum: the Tyco Electronics Copalum^{TM}, system, compression-bonding copper pigtails. Applied this to every single conductor except for grounding conductors is the most secure upgrade for aluminum branch circuit wiring, other than tearing it all out. My area has one authorized installer.

> **Inspection note**
> Copalum (TM)'s Listed uses still do *not* include pigtailing grounding conductors, and no such Listing was being sought, as the manufacturer confirmed near the end of 2009.

Other Solderless Connectors   There are two other plausible options, one of which I trust much less than I do Copalum^{TM}). Ideal Corporation's purple TwisterAlCu^{TM} is Listed for use with aluminum conductors, including use for grounding connections. However, it is widely misapplied; people often ignore the Listing limitations. I find a small, Listed setscrew connector with a separate entry for each conductor far more promising. Its copper aluminum connections have tested to the same standard as Copalum™ connections. Two differences may still favor Copalum™. First, the latter can boast decades of field experience. Second, its manufacturer maintains some control over who installs it, whereas the setscrew connectors can be misused, albeit perhaps not as wildly as are TwisterAlCu™ wirenuts.

Relay Switching

There are too many structured wiring/wireless control/mood preset/ integrated building systems for me to present an overview that would be current a year or two after this edition is published. Relay switching was installed more commonly in the past, for some of the same reasons more sophisticated control are installed now— much as PLCs are used where relays were used in the past for motor control. Just as happened with motor and controller offerings, where each manufacturer will advise what controllers should be used, building controls grew proprietary, once we left behind hard-wired multi-point switches and controls. With several competing communication protocols, whatever system is installed is the system that you need to understand.

Still, most older relay switching systems relied on regular Class 2 wiring. Cable sometimes suffers damage, and momentary-contact switches, relays, and transformers all wear out. I'd be extra careful: not just the transformers but also the relays present the hazard of 120 Volts. My sense, though, is that more and more often, older control systems are torn out. Some of the proprietary systems laid claim to particular market niches. This does not mean their principles of operation were, or are, anything other than powerline carrier, Class 2-wired relays, radio control, or, within line of sight, infrared. I haven't had to work on one for years.

Thermostats

As with relay-operated home controls, thermostats come, and came, in many versions. To replace a defective older thermostat, all I need is the number of wires connected to it from the wall, the operations that the thermostat controlled, and special features the customer wants. There's one important exception. Line-voltage thermostats are wired in series with the utilization equipment (normally heaters); no relays are involved. If an existing thermostat seems to be fed by BX or NM, I'll check for line voltage.

Another concern is that a two-pole line-voltage thermostat with a marked OFF position should disconnect all power from the heater it controls, but I have been told that some electricians have installed such thermostats so as to interrupt only one of the two energized legs. They work okay as controls, but leave the wiring potentially deadly.

Telephones

Telephones bring us back to low-voltage work. Communications cabling has progressed so quickly, and can be such a bear to untangle and trace out, that old phone lines warrant little effort or discussion, except where a customer finds a plain old telephone system perfectly adequate and wants it maintained. In such cases, old four-hole phone jacks can be replaced easily with modular units, and hard-wired telephones likewise. Old single line, four-wire cabling can be reconfigured easily to handle two lines.

Any complex system that needs significant repair or upgrading gets replaced, using a new design and new cables, starting with the backbone.

Example   One customer, occupying an old house, had a doorbell integrated into the phone system, by Totalplex. When I arrived, their "Doorbell Fon" didn't work, and I didn't spend a lot of time with it; it was just not worth the trouble.

I have three general final observations about low-voltage wiring. First, I can understand the inclination to skirt the requirement to remove old communications cables. However, I consider tagging an unused cable rather than removed, okay only when it is seriously being considered for reuse. Second, to minimize interference, I try to keep power wiring apart from data lines. Third, even if I am unfamiliar with a system, I particularly want to satisfy myself about one aspect of its installation: bonding/grounding.

Example   A fire burned a number of contiguous townhouses when a TV antenna cable fed an ungrounded monument and the cable passed quite near power wiring—without the systems being bonded together. A power surge arced from the overhead conductors to the antenna cable and from it to a grounded branch circuit—this last, inside flammable attic insulation.

Miscellaneous Rarities

I deal with a number of other odd items when working on old buildings. Unfamiliar overcurrent devices are an example of equipment that mostly calls for no special skills but does demand knowledge.

Old Breakers

One odd bird is the antiquated circuit breaker. Even a so-called "interchangeable" circuit breaker is not interchangeable with original

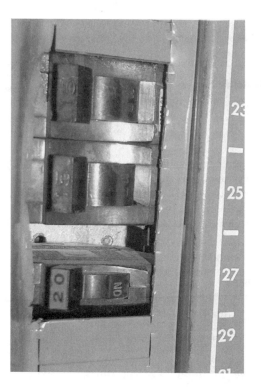

equipment manufacturer (OEM) devices unless Classified specifically for use in a panel, as evaluated by one of the NRTLs. Otherwise, tripping curves may not be coordinated—even when a CB comes off an assembly line in the same factory that makes suitably Listed or Classified breakers.

Another danger is that an inappropriately substituted breaker will allow more circuits in the panelboard than the manufacturer intended, or will create overcrowding in some other way. An inappropriate breaker also could have a different response to ambient temperature or other conditions. It may fit where the legitimate breaker of its rating or design would not, because of a rejection feature at that location in the panel. Or it may sort-of fit. Figure 6-34 (a close-up of Space 27 of the panel pictured in Fig. 6-35) is not the only case I've come across of a half-size breaker in a full-size space, with no filler plate to close the remaining opening. In this example, it is not a brand mismatch; I've come across some of these that looked as though it took a pry bar to fit them in and muscle the cover back in place.

I've seen many a breaker Listed for the panel in which it was installed—whether a new panel or an old one—but nonetheless installed in violation of its Listing. The device is installed where it is not permitted, according to the instructions glued in the panel. A common

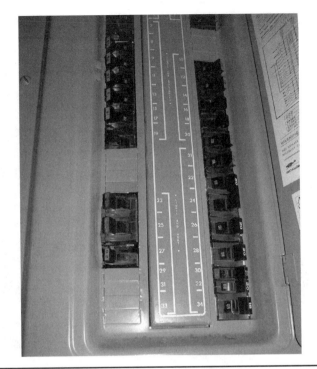

FIGURE 6-35 Note consecutive numbers above each other.

example is tandem breakers carelessly installed—sometimes in panels not permitting them, such as a 30-30 panel, as opposed to a 30-40—but sometimes just in breaker slots intended for single breakers, as in Fig. 6-36. Another common misapplication is installation of branch circuit breakers in spaces intended for MAINs. This is particularly common in split-bus panels.

When a circuit breaker goes bad—a rare occurrence, but one that I've run into—I want to substitute the correct replacement. Similarly, when I add a circuit to an old panelboard, I want to use a breaker Listed or Classified for the purpose. I urge customers to authorize the replacement of wrong breakers when I come upon them. At the least, I point the problem out to the customer. On an inspection, IFF a breaker mismatch appears to be preexisting, and isn't associated with the circuiting I'm there to inspect, I'm not going to fail the new work.

There are many panels in which the majority of the installed breakers violate the panel instructions. One consequence is that looking at what already is installed, or asking a customer what brand was used, doesn't provide a clue as to what should be there. I have to pull apart the wires and look at the legend, located with the schematic on the

---

Figure 6-36
Tandems don't
belong in this
panel.

door or inside the cabinet. Consider the strangeness of the numbering scheme and schematic of the oddball double-split bus FPE panel in Fig. 6-35, whose numbers indicate which spaces are for half-size breakers, and which are not.

There also are many panels, especially older ones, that at least do not present any difficulty in determining which brand of breaker is suitable. They simply require very different designs than the "interchangeable" plug-ins based on Westinghouse's early design. The CH and the QO have survived to be modern styles, for good reason. Zinscos have a much longer geometry than other styles, and they have their own means of attachment. For a brief period, Zinsco private-labeled these breakers to Arrow-Hart, as well. The cover of one distrusted old, split-bus version of the Zinsco panel is embossed with the service mark, "Magnetrip."

Another unusual style is the Pushmatic breaker, shown in context earlier, which has a single push-on, push-off pushbutton in place of a toggle handle. It is trip-free and indicating, with the button showing an "off" on its side, when it is either tripped or turned off manually. Figure 6-37 is a side view and Fig. 6-38 shows the mounting clip. Bulldog were the brand with Pushmatics, which was then taken over by ITE.

FIGURE 6-37 Pushmatics used bolted contact.

Federal Pacific manufactured the first plug-in circuit breaker. Their Stablok is said not to make good contact with the busbars. A more important consideration is that many of their breakers were well out of spec because they were manufactured utilizing badly worn dies, with the UL seal misapplied. This had been true of about one in ten tested on removal from working panels during recent research by Jesse Aronstein. His latest (Fall 2009) study, based on 830 Stabloks that were removed from service in early spring 2009, found defect rates ranging

FIGURE 6-38 The stab clips just secured mechanically.

from 20% of 1P, 15A to 70% of 2P, 40A or greater, for an overall fail rate of 28%. Stablok loadcenters no longer are manufactured under U.S. standards, but Listed Stabloks are made offshore, ETL-Listed.

Wadsworth made different unusual breakers (Fig. 6-39), whose ON-OFF buttons serve as a nice pop-up trip indicator–Braille-user friendly, perhaps.

Cutler Hammer was united with Challenger and Westinghouse and Bryant under one company, Eaton. One result is a bit of confusion. I have had quite a few customers ask me for a recommendation. If I tell them that Cutler-Hammer's CH breakers have an excellent reputation, I have to emphasize that Cutler-Hammer's BR breakers are a different line. Customers have also gone looking for QOs and come back with Homelines, not realizing the difference.

Some of the oddest breakers are noninterchangeable with a vengeance. In the early 1950s, apparently in response to problems that came to the attention of Chicago's Chief Inspector, for one and only *Code* cycle, I'm told that breakers were made that incorporated rejection features at least equivalent to those of Type S fuses. Thus, not only did a circuit breaker of one brand not fit into a panelboard of another brand, a 15-amp circuit breaker could not be replaced by a 20-amp of the same brand, thanks to panelboard geometry. In some cases, special tools were required to install and remove breakers. In mid-2006, I confirmed with Eaton/Cutler-Hammer technical support that it was quite acceptable to remove the bar that caused this from an early-1960s, 150-amp, split-bus residential panel so as to insert modern breakers. Removing the screws holding the bar or sawing it apart were two approaches suggested by the manufacturer's representative.

There are a few orphans, including Square D's short-lived 1980s Trilliants, which fit no other panels—any more than other breakers will fit the busbars in the plastic-cabinet Trilliant.

This is a good place to note that old breakers usually have 5 kAIC ratings. The 1962 NEC required AIC ratings on CBs, effective in 1965. Even residential occupancies nowadays often require higher AIC ratings because of utility design. I was troubled to hear that some utilities may choose transformers for residential neighborhoods indiscriminately. The medium voltage electrician who told me this said that, as a result, homes may be immediately downstream of transformers offering as much as 83 kA, more than many commercial occupancies. A lineworker confirmed this, noting that the actual fault energy available in a residence may hinge on the size of its drop or lateral.

Another worrying issue is this: a Schneider/Square D technical support engineer confided in early 2006 that their standard circuit breakers are good for at least 20 years, sometimes as long as 40. I've also been told that breakers decalibrate at rates depending on their internal construction. What about breakers in 50-year-old panels? I have tested the responses of existing circuit breakers in a suspect 25-year-old panel (Stabloks), to my personal satisfaction, but I can't do this for customers—I don't come onto a site with load banks. Even if I did, I'm still not a qualified tech.

I don't ignore the newer breakers with electronic circuitry. GFCI breakers are likely to have shorter lifespans because of the additional mechanisms. The AFCI breakers that Ideal's Brian Blanchette tested died after about 75 operations.

Finally, I don't want to ignore the three magic letters, CTL. If a panel is new enough to have the marking, "Type CTL," which stands for "Circuit Tandem Limiting," it is designed to reject breakers that don't belong in it. This could, for instance, mean any tandems in a 30-by-30 panel. It also could mean that only the lowest ten spaces accept piggybacks in a 30-by-40 panel. By no means is this found only in panels that have been very recently installed. A piggyback that is marked "not for use in Type CTL panels" could be used to bypass the rejection feature, illegally. Inspector Larry Griffith reminds me of how often this is done. Perhaps installers who do this don't understand what "CTL" signifies. Both QO designs have that label, as evident in Figs. 6-40 and 6-41.

FIGURE 6-40 A clear label shows this has no CTL feature.

Odd Fuses and Fuseboxes

I also have run into odd fuseboxes, although, no really odd fuses recently. In the past, yes.

Old Fuse Types

Old, rare fuses are less of a problem than old circuit breakers. To find homemade fuses of solder wire used commonly, for example, I'd

FIGURE 6-41 This style also says it's not Listed for CTL use.

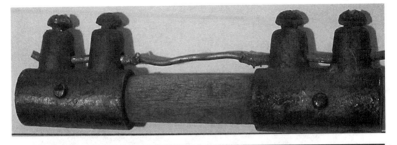

Figure 6-42   Fusible links used to be bare solder.

have to go back a century and more in the United States, or head over to some corners of Europe. I have come upon it here at least once (Fig. 6-42).

Substitutes are available for most old fuses. Often they represent a substantial improvement in function. I've replaced rebuildable cartridge fuses with one-time fuses of the same shape and ampere rating and the same or higher voltage rating and interrupting capacity. A fast-blow fuse can be replaced with a more common time-delay fuse, where appropriate. A fuse of odd ampacity might not be necessary. I'll replace a 4-amp branch-circuit plug fuse with a standard 15-amp no-tamper fustat, if it's not a dedicated circuit—or if it's safe for the device being protected. This evaluation is not complicated, unless the equipment's antique and no markings remain. I do make sure there's at least 14 AWG Cu wiring.

Combination Panels
Given the AIC ratings of fuses, panels that combined main fuses with breakers to feed branch circuits, such shown in Fig. 6-43, offered considerable benefit and are much simpler than alternatives. An interesting variation is shown in Fig. 6-44. This Murray panel offered circuit breakers to control some two-pole branch circuits, offering the advantage of fully deenergizing all the wiring in response to short or overload.

Energized-Front Fuseboxes
Some outdated fusebox designs can be dangerous. One kind of antique fusebox is the energized-front design. I've opened the door and not only the fuses but also the busbars were at my fingertips. (I'm talking about something different from a relatively new fusebox or cutout that lost or is losing its barrier, such as Fig. 6-45, a box which is too cheap to think twice about replacing, given time costs.) On rare occasions, I have considered cobbling together a safety barrier, voiding the ancient listing. As an inspector or installer, here are some factors I'd consider:

FIGURE 6-43   Main fuses protected CB branch OC.

- the barrier needs to be fairly rigid and to maintain its structural integrity;
- the barrier should be nonflammable, which rules out wood and some plastics;
- the barrier should not affect heat dissipation significantly, nor impinge on the wiring space orthe fuses and their terminals;
- the barrier should not itself pose a risk of injury, which rules out amateurishly cut and formed sheet metal;
- the barrier should not pose a risk of shorting.

In dealing with a energized-front panel, the least I'd do is to affix a permanent marking to the outside, warning users of the danger. The best answer, of course, is replacement—but this is not a call I normally can make for the customer, although they're easier to convince when I can report an imminent hazard, say because of corrosion or other damage that creates faults or high-resistance connections.

FIGURE 6-44   Fuse blocks and CB blocks mixed.

Asbestos

Another odd style in fuseboxes is the enclosure with asbestos insulation. There is no need for panic. Asbestos is a wonderful insulating material and it only constitutes a threat if breathed. Asbestos card (sheet asbestos) is not going to find its way into clothes or nose unless it's crumbling. Safely disposing of such a box, though, could pose a problem. Hazmat disposal becomes part of pricing the upgrade, which is going to have to be done sooner or later.

Switched Plug Fuses

Finally, some old fuseboxes have switches as well as fuses—plug fuses. They were made this way long in the past, and the switched-fuse design has returned—though not with plug fuses but with the latest minicartridges. Some had (and have) a master switch that disconnects the other fuses before the box is opened. This is a great safety factor, just like a modern fused safety switch or a main breaker panel. Others had a separate switch for each circuit. These switches can go bad; when they do, I have not managed to find replacement switches. This might mean it is upgrade time, at the very least to the extent of installing a subpanel.

Figure 6-45 A
broken protective
plate voids Listing.

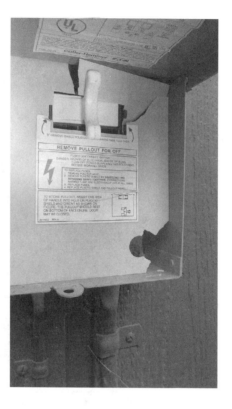

On the other hand, if switches for individual circuits fail in conducting mode—without adding resistance—this leaves a standard fusebox, in which circuits can be disconnected the usual way, by unscrewing fuses. I will add a sign noting that not all the switches are effective.

When the New Is Hazarded by the Old

Before leaving the discussion of odd circumstances with which I've dealt, it's worth touching on problems faced by very modern additions to old buildings. I've talked elsewhere about mismatches between new fluorescent lamp types and older ballasts. Here is one more example: delicate electronics. This category can include home computers; certainly it includes electronic point-of-sale systems.

There is nothing intrinsic to old buildings to cause difficulties for such equipment, but there are troublesome tendencies. I would not expect to commonly find triplen harmonics, except in commercial buildings that have had major equipment upgrades. However, I find many of overloaded circuits and outages, occasional arcing faults, and uncertain grounding. Too often circuits have been mixed together, as discussed earlier, yielding not only legitimate multiwire circuits but also circuits that should not be connected, yet that nevertheless share neutrals.

Then there is the issue of voltage drop. This is not an NEC issue. The tight voltage drop restrictions associated with sensitive electronic equipment involve separately derived systems.

Circuits having been confounded have tripped GFCIs and AFCIs, preventing their successful addition to detect ground faults or arcing. I still have managed to add safety devices and high-tech equipment. Sometimes I've put the GFCI closer to the ends of runs. Of course, dedicated circuits are the gold standard.

One last caution regarding existing high-technology equipment is that it might have been installed early in the learning curve. Recently produced GFCIs, AFCIs and electronic ballasts are more sturdy and more reliable than were earlier versions; in at least the more-sensitive situations I recommend replacing older ones. Other equipment too has changed for the better as manufacturers fielded complaints. Similarly, *Code* rules regarding new types of equipment evolve in response to field experience.

This applies to equipment as varied as photovoltaics and neon. One thing I always try to keep in mind is that older equipment, even when legally installed and not abused, may not have the same protections that have been required in recent years. Even when *Code* rules don't change, product standards may progress.

CHAPTER 7

Commercial, Industrial, and Institutional Wiring

Most of the challenges I talked about in earlier chapters can and do show up in commercial, institutional, and industrial installations, as well as residential. The unlabeled store panel in Fig. 7-1 is a too-typical example. Even the hazard resulting in the burnt connectors I've found, a plug in Fig. 7-2 and an octopus connector in Fig. 7-3, are not unknown. It had not occurred to me, but there is no reason for me not to have run into knob-and-tube wiring in institutional and other less-familiar settings. The examples in Figs. 7-4 and 7-5 show impressive workmanship. Still, I've run into certain conditions far more often in stores and institutions. (I haven't done factory work, unless the term "factory" is used very loosely). There are other conditions unique to nonresidential settings that are worth touching on, even though I can't offer first-person reports.

This chapter covers quite a bit of territory, but after the next few pages I talk mainly about two types of experience. I've been confronted with odd types of old equipment (at least in terms of power demand), equipment of a sort that I don't find in residences. Even more often, I'm asked to accommodate significant new demands on the old wiring, demands such as those posed by large refrigeration equipment. I've also been asked about adding lighting to an automotive body-and-paint shop that had been in existence forever. The owner had no concept of the need to use adequate ventilation in potentially hazardous locations.

Even more frequently, I have encountered a parallel to the residential "our house just isn't big enough." The storekeeper says, "To stay in business, we have to keep on hand the products our customers want; we have no other place to put them. And no, this isn't ideal, but we haven't burned down yet." As a result of this type of thinking,

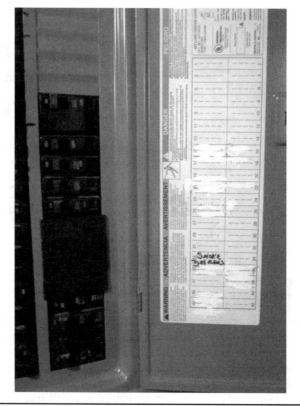

FIGURE 7-1   A directory must be replaced.

FIGURE 7-2   Heat damage can show up in any occupancy.

FIGURE 7-3 An overloaded device will overheat nearby equipment.

Figure 7-6 shows a token sign, which I zero in on in Fig. 7-7. Figure 7-8 shows wall-mounted magazine racks that block panel access. I'd hate to remove the cover (Fig. 7-9) for energized testing while the rack remains in place.

One manager tried very hard to make his small store safe—but didn't have much of a budget to let him hire electricians. He did heed the advice I managed to offer on the occasions when he was forced to

FIGURE 7-4 Our military used classy K&T work.

FIGURE 7-5   Some tubes even contained loom.

FIGURE 7-6 Signs
don't keep space
clear.

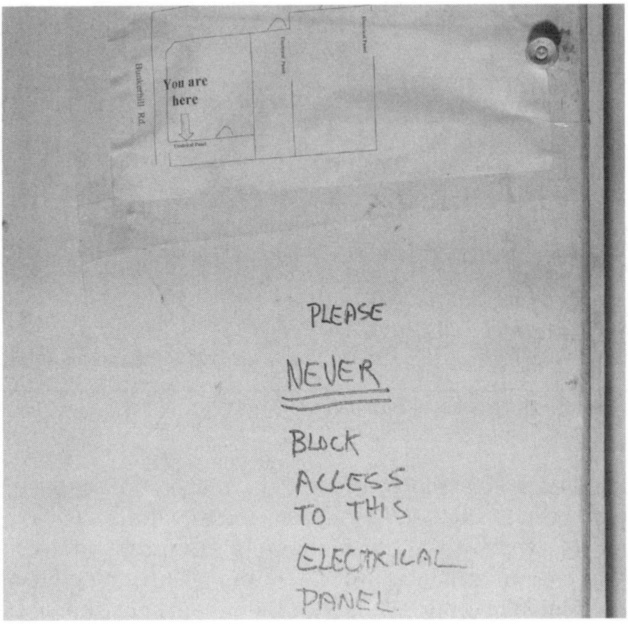

FIGURE 7-7 Even strongly worded signs are not enough.

hire me for troubleshooting, with the result shown in Fig. 7-10. Much of the equipment was dated. The service panel in Fig. 7-11 had a major problem, by current standards. Yet, as the closeup in Fig. 7-12 shows, the MAIN's orientation was legal when installed. Only in the 1984 NEC were panelboards added to switchboards in requiring UP to be ON.

FIGURE 7-8 Unscrew the rack; remove the cover unobstructed.

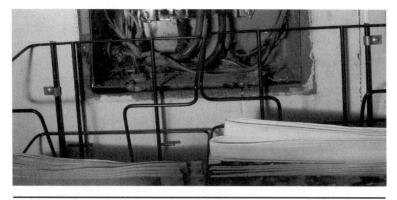

FIGURE 7-9 Sliding the cover out instead is nuts!

Like the stapling shown in Fig. 7-13, much or all of the original work had been nicely done. Additions to the wiring, such as shown in Fig. 7-14, were flawed. Even attempts to support and secure cables, such as you see in Figs. 7-15, 7-16, and 7-17, didn't quite comply with Section 300.11, nor 110.3(b)—although, despite the violation, I suspect he improved on what he inherited. Many runs were overlooked, throughout the basement. In one section of the room, Fig. 7-18 ignored Section 300.11(a); in another, Fig. 7-19. He looked right past some cables, as evidenced by his effort to create order (Fig. 7-20).

I will return to smaller commercial operations later. The bulk of this chapter departs radically from the course of the rest of this book,

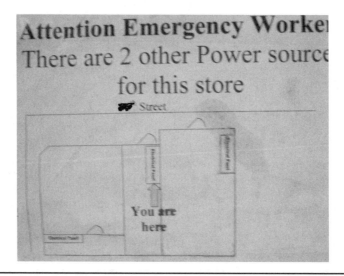

FIGURE 7-10 This old store had three services.

FIGURE 7-11 The MAIN-on position was down.

and talks theory; it also talks about what people whom I respect have done, when faced with situations I have not dealt with myself.

Special Service and Tweaking Power

I start by looking at situations where there is a mismatch between the utility's service and the type of power required by the occupancy's

FIGURE 7-12 It was legal here in 1966.

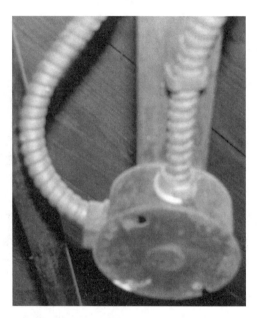

FIGURE 7-13 Old BX was stapled, new—strained—not.

FIGURE 7-14   Old BX was stapled; new, tied to pipe.

FIGURE 7-15   Cable was supported by cable.

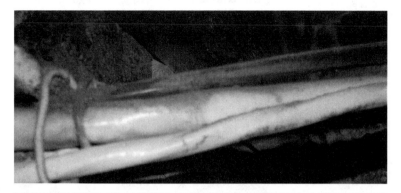

FIGURE 7-16   Cable also was hung from enameled RMC.

FIGURE 7-17 BX
was tidily strapped,
but with ty-raps.

FIGURE 7-18 Even some cable on wood got no staple.

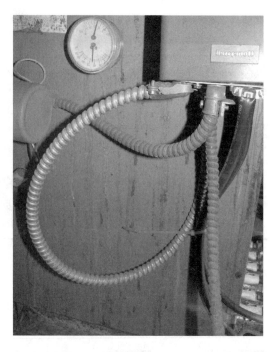

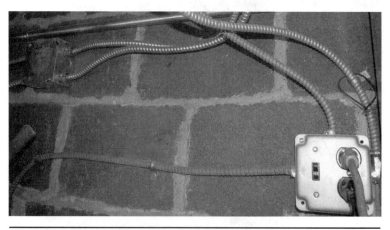

FIGURE 7-19 Only the tarnished BX has straps.

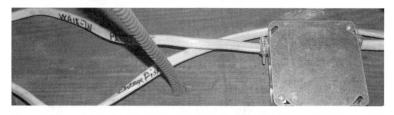

FIGURE 7-20 Extra labels are great, but support's critical.

equipment. This will lead naturally to some cases of dealing with old, unusual, equipment.

First, there are a number of ways to finesse the supplied power to create the equivalent of a different class of service.

Not Quite the Right Voltage

Buck-boost transformers are a common and simple example of how voltage is adjusted. Autotransformers commonly are used to step up and down between, say, 208 and 240 volts, respectively.

There are systems, usually in industrial settings, utilizing 208 volt, single phase, obtained between the high ("red" or "stinger") leg of three-phase 240-volt delta wiring and the grounded midpoint between the other two phases. This 208-volt power exceeds 150 volt to ground, unlike that obtained from 208Y 120-volt wiring.

Phase and Frequency Differences

I have not myself worked with the odd old systems discussed in the next third of this chapter. I do have full faith in the people who were kind enough to discuss them with me, but there is a huge gap between understanding systems in principle and mastering nuts-and-bolts application. Because I have not worked with synthesized three-phase, DC, two-phase, or 25-Hz power, my reason for describing each of them is to help other readers avoid getting caught flat-footed if they run into it. I'm more than willing to say, "I don't have any experience with this." What I hate to say is, "I'm afraid you must be mistaken" when the person calling me in fact knows their system well enough that they are providing an accurate description.

Later, as I talk about upgrading more modern motorized equipment, about commercial and industrial lighting, and about improving energy efficiency, this chapter will move toward the more concrete. The stories of poorly planned HVAC upgrades are painfully graphic. The extensive discussions of errors in maintaining terminations and bonding mix theory and practice sufficiently, I trust, to satisfy both the engineer and the maintenance electrician. For now, though, those readers who care for the diversion get to read about increasingly rare distribution systems and, by implication, utilization equipment.

DC I'll start with unfamiliar frequencies. The simplest frequency supplied is current with no intentional alternation at all: DC. This was the very first electric power system.

If a customer was one of the few still using utility-supplied DC, I can't believe they won't tell me. Here, though, as everywhere else, I'd use my tester. One minor problem is I've had plenty of customers talk to me about their "positive" and their "negative" wires, when they simply meant their energized and neutral conductors. Even with DC service, this is not an equivalent description. Utility-supplied DC, generally 120/240, not only included a positive and a negative conductor, but also a neutral.

Why consider DC? Existing equipment is one reason. Besides, DC motors are very dependable. John Esemplare of New York City's Con Ed has seen at least one Westinghouse DC motor in regular use that looked like new after more than three-quarters of a century. Electric rail systems in the Northeast Corridor still distribute DC power, because they own so many DC locomotives; they invert it in the more-modern locomotives that run on AC. The Washington, D.C. area's subway system, Metro, takes utility AC power at 13.8 kV delta to traction-power substations that convert it to 500 VAC, and then, using rectifiers with as many as 18 poles, to 750 VDC. Their latest transformers take the 13.8 kVAC and send it to the rectifiers on their secondary side through both three wye-connected and three delta-connected coils—six buses—to reduce damaging ripple. They feed the rails two-wire DC, with the rails serving as the negative-return conductors.

As of the end of the twentieth century, aside from traction power, utility-supplied DC served a few, rare industrial customers. Also, some older buildings in large cities had elevators or other equipment run by DC motors. Sooner or later, all these will be gone.

The arguments for and against DC are not relevant to one point: I know that it is still around and that I am not familiar with it. But I do have colleagues like Ed Bihlear, who has worked on it. I wouldn't try to wing it.

Single- to Polyphase   Almost all residential wiring is single-phase. Even in multifamily dwellings served by three-phase, four-wire, 208Y 120-volt power, the wiring within each occupancy normally is single-phase. I have had only one residential customer whose single-family home—not small, but not enormous either—had three-phase service. Where three-phase power is not available or convenient, a couple of expedients can work.

Synthesized Three-Phase   There are two ways to convert to polyphase power: rotary and static. The first employs either a rotary phase converter or a motor generator.

A rotary converter is, in essence, a three-phase motor designed to run on single-phase power; its connection to the power system parallels a regular three-phase motor or motors that are part of utilization equipment. A motor generator is, essentially, a single-phase motor turning the shaft of a true three-phase generator. The latter tends to be rather more expensive, though its three-phase output is more reliable than that produced by phase converters. With a motor generator, there is no synthesized phase to be wary of using to feed single-phase loads, even incidental ones.

The second basic approach, the static converter, simulates a third phase using capacitors—either with capacitor-relay type synthesis or with a solid-state device that produces three-phase power by rectifying single-phase and synthesizing the desired wave forms. This approach, applied injudiciously, can badly affect nonresistive loads in the rest of the system by generating harmonics.

Even with this drawback, I've been told, in some situations three-phase motors can be more efficient and durable than single-phase. Phase converters have been used, for instance, to equip sewage lift stations with three-phase pumps, when three-phase power was not conveniently available.

In the past, delta breakers were used to bring a third phase in from upstream to feed one three-phase motor load from a panel that otherwise had only single-phase loads. Only replacement of existing ones is legal, because if the normal source of power to the single-phase panel went off, it could be backfed by that three-phase motor load, because the third phase entering it from upstream of the single-phase panel kept the motor turning.

More Unfamiliar Frequencies From the issue of single- versus three-phase power, I will move on to more exotic types of distribution, or at least utilization.

Putting aside DC, 60-Hz power may be today's old familiar standard, but it has not been the standard forever.

50 Hz As of the very end of the twentieth century, 50-Hz power was available from some U.S. distribution systems. The standard in Europe is still 50 Hz. Because of this, companies that import European machinery sometimes need 50-Hz power. If the utility doesn't supply this—and why should it?—the company needs to convert 60 Hz. I received an inspection call at the end of 2009 from a healthcare manufacturing facility well under a decade old that uses 50 Hz for this reason.

25 Hz Even 25-Hz power still was supplied to some large old industrial customers all around this country. I have been assured that the 25-Hz frequency is terrific for incremental control of motors such as those used for elevators. As recently as 1992, the Firing Circuits Company of Norwalk, Connecticut, would get an occasional inquiry for 25-Hz motor starters. For twenty-first century 25-Hz experience, though, there's a better place to look.

Example: The Railroads According to Bob (Victor) Verhelle, Amtrak's deputy chief traction engineer, there are a number of reasons to retain 25 Hz equipment. I'll go into the benefits after describing the system a little further.

The 25-Hz utilization frequency is derived from utility-supplied 60-Hz power. Conversion originally relied on motor generators, but these incur greater maintenance costs than do later systems. To serve loads below 500 W, conversion once relied on vacuum tubes, later, on diodes, and then SCRs. As of early 2006, they used three types of converters and this is not likely to change. Some rectified 60-Hz power into DC, cleaned it up, and synthesized the 25 Hz, some used rotary converters, and some used solid-state cycle modulation.

Because 25 Hz offers some of the torque benefits of DC and some of the transmission efficiency of 60 Hz, it became the standard at one point.

Another advantage to this compromise has to do with electromagnetic interference (EMI). This concern favors the use of DC and, to a lesser extent, any lower frequency, such as 25 Hz, when passing through cities. Moreover, because the rail is fed off a center-tapped autotransformer from the catenary (the high-voltage cable running overhead), current flows through it in the opposite direction to catenary power, causing the fields to cancel.

Amtrak is a powerful example of an end-user that is large enough to maintain substations supplying whatever frequencies it needs. A complete changeover, however, was not feasible.

Amtrak explored total conversion to 60 Hz in the early 1990s and determined that the new substations would have cost a cool billion dollars. Instead, a 25-Hz traction-power network is supplied from six utility locations. At five of the locations, three-phase 60 Hz is converted to single-phase 25 Hz for transmission over their 138-kV network. Traction-power transformers at substations send 12 kV to the pantographs of the electric trains. The pantographs collect the power from the catenary for the trains to operate on. Elsewhere, automatic tap changers let these same trains run on 12.5 or 25 kV one-phase 60 VAC provided from either 138- or 115-kV taps from the utilities. This includes their newest trains and their 1970s trains, which they gave new engines in the 1990s.

Two Phase   It is time to return to systems that are unusual not because of frequency but because of phase. Two-phase systems, some three-wire, some four-wire, and some five-wire, still can be found, primarily because people still have motors designed to work on two-phase power. In two-phase systems, the phases are 90° apart, unlike three-phase, where they are separated by 120°. Multiple companies were still advertising Scott-T transformers, which are used to synthesize two phase, in late-Spring 2010, as this edition was being edited.

The New Jersey utility, Public Service Electric and Gas Company, would offer new two-phase service, to customers who wished it, at least into the 1950s. Subsequently, they supplied it to customers who had existing two-phase equipment, converting to two phase at pole-top. As of the end of 2001, at least, they still had customers in some older cities operating two-phase elevators or manufacturing equipment. Often, when they changed transmission voltage, say from 4 to 13 kV, they provided assistance to their customers to convert their equipment to three-phase. They utilized a Scott-T connection to supply two-phase, five-wire 400-amp 240/120 volt power to one last customer in Bayonne until mid-1996, when the customer's building was demolished.

Here is what a PSE&G official relayed in 2005:

> Two-phase power service was as the gentleman indicated, an earlier way to serve power load. It is no longer provided to <u>new</u> [emphasis mine] customers and has not been for at least 40 years. Often we assist customers with older facilities in converting their internal equipment to three phase if we are doing work in the area such as a conversion from 4 to 13 kV. But there are still some customers in our older cities that have two phase service that supply load such as elevators and older manufacturing equipment. The difference for PS is that we use different pole top transformers to step the primary voltage down to the secondary two phase to feed the customer. If their equipment fails they usually replace with three phase.

While rare, two-phase equipment is something a few readers just conceivably may need to evaluate, inspect, or troubleshoot. The 2008 NEC recognizes three- and four-wire two phase, in Table 430.249. If I were inspecting or performing plan review, that's where I'd go for the current values. Even when no longer utility-supplied, two-phase equipment, like DC equipment, continues in use. As of late 2001, Philadelphia Electric Company (PECO) also still supplied some two-phase power. The Niagara Mohawk utility in upstate New York sold the transformers it had used to derive two phase to some large customers, owners of weaving mills, in the late 1960s. As of at least 1997, the transformers still were in use to operate the many two-phase motors the mills owned. Apparently these motors are very durable. Two-phase motors were still being manufactured in 2010 .

Nonetheless, because of their relative rarity, when such motors burn out, most customers should be happy to replace a two-phase motor with a three-phase system. Given that three-phase is more efficient, there would seem to be no reason not to make that change rather than to find someone to rebuild two-phase motors. A potential problem when there are many two-phase motors, and only some are replaced, is that the customer could end up having two- and three-phase motors working side-by-side, requiring separate feeders.

I talked about this to the representative of one interstate, for-profit, utility in early 2006. He told me that they have gotten much more competitive in recent years. What this came down to is that if a customer asked them for two-phase power, they probably sent the customer off

to find another supplier. It was not worth their while to accommodate such unusual requests; that's what "more competitive" meant to
them.

I just might be called upon to maintain, inspect, or evaluate a two-
phase supply, but the chance is vanishingly small. Motors are not the
only use. Apparently, two-phase power has some advantage in arc
welding.

Addressing an Awkward Class of Service

Why would people work with systems that have been generally
phased out? Money savings are one reason to fiddle with the nature
of incoming power. It might seem far more sensible to order the right
equipment and the right utility service, so that they match each other
from the start (or at least from the time you appear on the scene).
This option is not always available. It may be far more expensive, at
least in the short term, than one of these alternate means of supplying
equipment with the type of power it requires. If the approaches in
place violate neither manufacturers' (and Listing agencies') specifications nor the NEC, as an electrician I still could choose not to maintain
them. However, as an inspector I would want to understand enough
to evaluate their installation for *Code* compliance. As a consultant, I
would not want to dismiss them out of hand because of unfamiliarity.
In either case, if I felt totally lost I'd kick the problem upstairs, or to
someone more knowledgeable, if I could find such a person.

> **Inspection pointer**
> I have listened to and talked with colleagues who have worked
> on medium-voltage systems, on DC, on systems requiring extremely
> high reliability—"six nines." This means a reliability of 0.999999, or
> 99.9999% reliable delivery. I think I followed the explanations, but this
> does not make me a competent person to inspect the equipment. Each
> one of them has said, though, that the exotic nature of their work is
> that they have had to teach the inspectors when they came.

Example   As they were connecting up PLC equipment at one substa-
tion, Ed Bihlear reported, the manufacturer's engineers discovered
that the factory had hooked things up so that 130 volt fed the 22 AWG
communications ground. The current melted the wire, and the field
engineers had all the PLC remotes replaced, as they had been subjected
to this out-of-spec power. After this, I can believe the crew carefully
checked the rest of the setup extremely carefully. Even so, Ed, who
was responsible for commissioning the new system, found enough
problems to fill ten pages. Other sites have yielded more problems
than that to his experienced checking.

It can be argued that it would be my responsibility to point out options for at least replacing systems such as two-phase motors, when they could save the customer money. The calculation involved in determining this, however, can be complex. Comparisons of life-cycle costs have to take into account at least the following considerations:

- Out-of-pocket costs, including the difference in value between money spent now and the same money invested ("the time value of money")
- The risks associated with having two different types of power in a facility
- The cost of down time required for the changes versus the cost of unreliable operation and the difference in cost between scheduled and unscheduled down time
- The training costs associated with making any change versus the costs to train new personnel to deal with unfamiliar types of power
- The present and future costs of power, both in terms of kilowatt-hours and of peak demand and power factor penalties versus cost penalties for obtaining exotic power— including the inefficiencies associated with converters
- Differences in equipment life span
- Costs associated with differing maintenance requirements
- Costs of having to increase electrical system capacity, versus the savings and expenses associated with making do with what's in place

Another consideration to keep in mind with regard to unfamiliar solutions is whether they will deceive users into misapplication. Burning out a few motors, or starters, or even lights, on the wrong voltage can wipe out initial savings very quickly. This does not even take into account the "we can't get this to work" callbacks.

Upgrading Class of Service

Because of all this complication and the inherent inefficiency associated with phase conversion, it is worth backpedaling just a bit. If I were called in because phase-conversion equipment—or the associated utilization equipment—were reaching end-of-life or rebuild time, I would seriously research the possibilities for upgrading the building's service. This would be true especially if the present electrical system were deteriorated or bordered on the inadequate. There might be advantages to checking out the possibility of obtaining a different class of service. The serving utility often will have changed policies,

making this more attractive than in the past. Alternately, new, perhaps more efficient equipment that matches presently available power
without the need for phase conversion may prove to be a cost-effective
investment.

On the other hand, if three-phase power is not readily available
anyway, it is likely that the serving utility would levy a per-foot charge
to bring a four-wire wye line to the facility, and perhaps a per-pole
charge if new poles need to be set. (This assumes there is no problem
with easements.)

"Plain" Old Motor Replacement

The remainder of this chapter sets aside voltage and frequency, and
deals with less exotic systems. The next section continues to look at
motors and motorized equipment, but in less unusual situations. What
I find least unusual, unfortunately, is that motorized equipment is
installed without qualified people involved. Figure 7-21 is an example,

FIGURE 7-21   HVAC guy probably overfused, did misuse the lug.

where the line side, field-wired, shows 12 AWG wire and a 60-amp breaker; something's fishy.

> **Key note**
> Many snares that could attend to simply replacing a less exotic motor also apply to replacing an older motor that relied on a different type of power.

Regardless of a building's age, when called in to deal with a sick or dead motor, after I check incoming power I usually call a motor shop. Some equipment makes sense to replace altogether.

I can call my local motor shop about a motor, or bring it in—in the sizes and types I've dealt with so far, anyway—and they will tell me what makes sense to do, and whether there are any down sides to asking for a matched replacement.

Of course, if the equipment use, layout, or environment has changed, some redesign may be needed. Improvements may be possible. I don't know a great deal about motors, but I know enough not to think I can simply replace an existing, old motor with a more efficient modern motor without checking any number of factors, from inrush to physical layout to the kind of power disturbances that may be present to the ones that might be created by new controllers.

One common characteristic that has changed is that new standard motors, in the early twenty first century, are 75° rated.

Exotic Motors

There are exceptions to some of the rules we know about, for instance, the relation between operating efficiency, inrush current, and starting torque. One is the written-pole motor. There are others. Because I haven't worked with them, I will leave my mention of it at this. If I were faced with problems in replacing an old motor, like any reader, I would need to do research specific to the situation.

Matching the Power to the Load

Finally, here are two facts familiar to many electricians who have worked with motors. First, while it can be wise to install heavier services and feeders to allow room for growth, this principle does not apply to choosing the motors themselves. I would want to avoid loading a motor to capacity, relying on its service factor for slack. However, oversizing a motor by one-third or more results in much lower efficiency than does matching it more closely to its load.

Second, when installing it, I'd make sure the replacement motor turns in the right direction. Overlooking that one test can result in a silly—and potentially painful—call back. Even if I'm the inspector or consultant rather than the installer, if I'm involved in commissioning I'd be inclined to ask for this demonstration.

Leaving It Functioning and Safe

I will carry the principle further. I dare not assume that I can get motors repaired or replaced and leave it at that. I may be responsible to see that the installation provides a modern level of at least of personnel protection when I restore it to operation. I'm talking about switch enclosures, conductor protection, and disconnects and controllers. This concern is always present when the equipment looks kludged, rather than just old. In other cases, this depends on the AHJ and on how much change I've introduced into the electrical system. Some very hefty OSHA penalties have been assessed for such violations. I haven't personally dealt with high-voltage equipment, so I will not go further into the details of requirements such as warning signs. I also have no experience selecting options for safety compliance, such as alarms, failsafes, and automatic disconnects.

Panelboard arrangements
Since most three-phase panels are in commercial or industrial establishments, or at least large institutions, this is the best place to bring up the issue of changes in phase arrangement. As mentioned under 'Not Quite the Right Voltage,' a delta service with center-tapped phases has one phase furthest from ground. This must be connected to the "B" or middle busbar. That requirement only came in as the 1975 NEC was adopted. Most commonly, the phase further from ground will be the third, "C" leg in pre-1975 systems. The exception to the current requirement, where the serving utility connected a meter differently, is even more recent. This makes testing especially important.

Lighting Upgrades

So much for motors. It is time to look at lighting systems.

By the time this edition has been in print for a while, light-emitting diodes (LEDs) may indeed be generally competitive in life-cycle cost and in lighting with fluorescents. I keep checking the reports that are posted every few months at http://www1.eere.energy.gov/buildings/ssl/reports.html. (*Consumer Reports* eat your heart out.) As of Spring 2010, their latest (October 2009) test summary showed T8 fluorescents in troffers beat the LEDs for lumen-per-watt efficacy. Other applications of LEDs already meet or exceed fluorescents' efficacy, according to their tests. They do point out that LEDs tend to produce more glare than light sources that are relatively diffuse. Heat dissipation and life-cycle costs are separate factors to consider. Too much current can destroy an LED, as can water or too much heat. If known voltage fluctuations might overwhelm the power supply, an LED might not be the best choice.

There's also the fact that for efficient operation, LEDs require their own drivers and, to quote DOE, "... Buyers, however, should be wary that rewiring a troffer to bypass the ballast may jeopardize the UL certification (or other testing certification) of the troffer . . . ." Of course, just as with CFLs, an LED or OLED (Organic Light-Emitting Diode) retrofit kit could be Classified for the use. In such a case, as an inspector I'd have no gripe if it were installed within its Listing restrictions; as a contractor I'd be inclined to challenge any inspector who came down on the use, to find out whether there was some other cause for objection.

Exit Signs

Replacing exit signs, or relamping them with appropriate Listed retrofit kits, can be a simple source of savings. Because the signs must shine continuously, a considerable amount of energy can be saved by replacing incandescent lamps with fluorescent, and doing so also extends relamping intervals considerably. Retrofit kits utilizing LEDs offer even greater labor savings.

Fluorescent Fixture Inserts

Moving on to linear fluorescents, even when there is no change in wiring, any retrofit kit, like other Classified equipment, must be applied in accordance with its Listing. Such is the case with specular (extra-shiny) reflectors, designed to reduce the amount of light lost before it even leaves a fluorescent fixture. A NRTL evaluates and classifies a reflector kit for fluorescent fixtures as one of two types. Type I is a simple reflector, which either sticks to the fixture's existing reflector or else replaces the original reflector. Type II is more sophisticated. Besides increasing reflectance, it will often reduce the number of lamps; doing so may require reballasting or changing lamp locations in the fixtures. A word from the field: One of the down sides is that it can take me some force to remove these from linear fluorescents, or to reinsert them–and the reflectors' edges can be <u>sharp</u>.

The way to proceed will vary. Some of these fittings are Classified for use with specific fixtures within particular manufacturers' product lines. Others have been investigated for use, say, with all 2'-by-4' troffers.

Some jurisdictions require permits for such retrofits; others do not. As an inspector, I wouldn't fuss about it. However, if it were part of a job I was inspecting, I would want to confirm its Classification for use with the particular luminaires while I was there. Similarly, on repairing a fluorescent I think it's a great idea to install local disconnects in both conductors entering. However, I can only see requiring them on new installations.

The SWD requirement

Before I leave the subject of lighting, here's a specific issue I have learned to be alert to. If the old lighting system is incandescent, controlled directly by circuit breakers, there's a Listing issue associated with changing over to fluorescent lighting without upgrading the controls. (This can be resolved simply by installing a downstream switch.) Even re-lamping with compact fluorescents (CFLs) brings in the requirement that the circuit breakers used as lighting control switches be SWD-rated, in the opinion of many, including *Code* expert Charles Eldridge. Modern 15- and 20-amp CBs carry the marking. Old circuit breakers may or may not have this rating. Certainly I would not expect that any of the no-longer-Listed product lines would have SWD ratings. For instance, I checked several Bulldogs: nothing. Offshore-manufactured CBs, ETL-Listed, for brands such as Zinsco and FPE, apparently are all manufactured in compliance with the latest standards, including SWD ratings.

Square D couldn't tell me precisely when this changed in their branch circuit CBs. However, the best estimate Schneider-Square D's Technical Support crew could give me, in mid-November 2009, was 30–40 years. This means that this issue does not arise with Homelines, or with their electronic breakers such as GFCIs. However, a QO that might have been installed before 1970—even before 1980—would be worth checking for an SWD rating. Square D's XO breakers were manufactured only up to the late 1960s; they won't have it. This is aside from the fact that over that long a period, CB calibration could easily drift, as mentioned in the previous chapter. As for Cutler-Hammer, Bryant, and Westinghouse, I asked a representative of Eaton Technical Support, the same afternoon I contacted Schneider-Square D. He frankly told me that Eaton Technical has no idea. I left messages with Siemens/ITE and GE, but did not hear back.

This means that we can't rely on dates. Fortunately, if I can make out the small print, the breakers themselves will tell me.

CFLs

I think of compact fluorescents, especially the low-end retrofitted screw-ins, as mainly employed in residences. Then I walk into an old nonresidential occupancy with dozens of incandescent fixtures, if not more.

That's why I'm addressing them in this chapter. There are several additional issues related to functionality and to Listing. The first functionality-related aspect is this: like other fluorescents, CFLs may be temperature-sensitive, in the opposite direction from LEDs. The second is that their longevity is reduced by brief ON-OFF cycles, as may be the case with lights in storage areas. This limitation has been mitigated in newer designs. The third concern is that the ones designed for non-commercial use have low power factor.

When someone oblivious to Listing issues has retrofit incandescent luminaires by relamping with CFLs, especially in an older structure, I want to glance at two elements. One, has anything about the luminaire been tortured to make the lamp fit in the fixture? This can be dangerous and even the most compact ones simply don't fit in some luminaires. Two, does the CFL have any Listing restrictions—they will be printed right on the unit, that forbid its use in the location, meaning the environment or orientation? I'm not just talking about the fact that most CFLs—basically, the standard screw-ins that don't carry a premium price to match a premium capability—are not designed for dimming.

For a long time, I understood that a CFL was unsuited for luminaires whose markings specified incandescents. However, CFLs and LEDs whose configuration matches that of less-efficient light sources are essentially Classified for use in luminaires designed for those sources, according to UL.

Laypersons can read some instructions over-literally. I'am glad for them that I'm the person who's been called on, rather than someone greedier. There are customers who assume that if a CFL states that its light is equivalent to that of a 100-W incandescent, it necessarily will be. Others, far more confused, assume that a 13-W CFL that claims it is equivalent to a 60-W incandescent is the brightest that may be installed in a luminaire marked, "Maximum 60-W."

Easy Improvements in Lighting Efficiency

The concept of evaluating the efficiency of lighting systems is quite old, as I found demonstrated in a 1917 table for designing lighting by taking into account lumens/watt. Many systems still offer plenty of opportunity for improvement. Most of the changes proposed in the name of energy-efficient lighting are quite blind to the age of the building or of its electrical system. These include the following:

- Occupancy sensors.
- Time controls, including daylight sensors and time clocks.
- Delamping.
- Relamping, when fluorescent lamps are aging.
- Installation of more efficient reflectors.
- Fixture cleaning (simple, but not to be discounted!).
- Fixture lowering.

- Purchasing higher efficiency or reduced wattage lamps or ballasts of the same type. As linear fluorescents have grown slimmer, for example, shrinking from T12s to T8s to T5s, the increased closeness of the arcs to the phosphors has made their use of power more effective. Especially when it's time to change ballasts, this can make a whole lot of sense. The only complication can be the need to stock a mix of different lamp types, such as F32s and F40s, if only some of the luminaires are changed over. Even incandescents have enjoyed improvements, especially in filament design.

So long as I encounter no deteriorated equipment or materials, there is no difference between making such changes in an old building and in a fairly new one. Of course, the old building's conductors are likely to be rated for 60°C, but while this is an issue when I change luminaires, it should not affect replacement of lamps or controls.

When System Age Needs to Be Considered in Pursuing Efficiency

Most, but not all, improvements in efficiency not only will save energy, but will also extend the life of the wiring system. Certain changes, though, can create problems for old wiring and service equipment. Both observations are based on the far greater likelihood I have seen for old wiring to be loaded nearer its ampacity in a commercial setting. On the one hand, this means that reducing the load can help by lowering the operating temperature of the wiring system. On the other hand, because commercial and industrial installations tend to use multiwire circuits for lighting, I could run into the problem of overloading neutrals by changing to nonlinear loads in sufficient quantity. As I mentioned elsewhere, in some old systems a neutral could carry the return current of not just two or three conductors but of more—either by antique design or inadvertently. It's something I consider.

As a side note, I know I could find open/cleat-mounted, knob-and-tube wiring utilizing unprotected conductors in noncombustible locations in some industrial settings. Without dielectric insulation to deteriorate, these could essentially last forever, even at the elevated ampacities that were permitted at one time. If conductive or combustible materials were moved into spaces where, say, 12 AWG wires were protected—legally—at 30 amps, serious overheating could result from the addition of thermal insulation. The worst scenario would be where blown-in insulation became combustible. The fireproofing on cellulose insulation, for example, can deteriorate in well under a decade.

Example At an International Association of Electrical Inspectors (IAEI) *Code* workshop, I saw slides of saddle burns on joists, burns specifically where overheated conductors lay across them in an attic space. The conductors were overloaded, they were spliced inappropriately and they were protected worse. Even so, they might not have caused a fire had the space over them not been retrofitted with thick thermal insulation that trapped their heat. Jim Rogers, the electrical inspector and forensic expert presenting the material, notes that adding the thermal blanket violated restrictions in Articles 334 and 338; I would be inclined to add Section 110.11—which I find in place as Section 1110 in the 1950s, and part of Section 206 in earler editions.

Other Risks Associated with Old, Nonresidential Buildings

So much for equipment protection. In Chapter 2, "Hazards and Benefits," I began to talk about personal safety problems associated with old buildings. Old industrial and commercial buildings probably pose greater risks of exposure to hazardous materials than do old residences. From asbestos and rock wool to solvents and process chemicals, there can be many toxic challenges to worker health on the job– and usually no Material Safety Data Sheets. If I'm not sure what I'm dealing with, I am inclined to assume the worst until I find someone I trust thoroughly who does know for sure. This is not just somebody who tells me, "Hey, it's okay."

One serious danger found especially, but not exclusively, at industrial sites is that over the years the serving utility may have upgraded its transformers and lines. This easily can increase the available fault current at equipment that already carries enough energy for serious arc flash danger to a level that exceeds installation ratings, even though the initial installation did not include any misapplications. I certainly don't expect the serving utility to automatically notify customers of changes.

Another danger is that lugs may have been overtorqued or subjected to repeated torquing. The problems that practice can create will be discussed later. For both reasons, some very knowledgeable persons recommend standing off to the side, "out of the line of fire," when operating any disconnects, to provide partial—but only very partial—protection in case of low-level arc flash or blast. This is no substitute for PPE. Basically, I use this to get myself out of the way of possible sparks, or very slightly worse. I use this when operating a disconnect in a low-power environment, such as the average SFR fed by relatively small gauge utility conductors, when, for example, I'm not sure of the integrity of the enclosure. Many old disconnects have missing or damaged barriers.

Examples: Ill-Considered Upgrades   Another kind of risk associated with modernizing older industrial buildings is that if I try to update without considering all aspects of the customer's needs, improvements can backfire—and this can mean that, at the least, I could have to eat the expense of doing things over. Three examples follow.

In the first edition I wrote about ill-considered efficiency upgrades to HVAC systems in three buildings. One corrected a system that had kept chocolates wastefully cold. Unfortunately, the replacement allowed the chocolates to get just warm enough, periodically, for quality changes to result, making them unsalable, even though they technically remained wholesome. In the second warehouse, the goods had been kept much cooler than was warranted by product characteristics and length of storage. However, as the temperature rose, vermin attacked. The old building's envelope was too porous to keep them out. In the third case, a condensation problem was solved with an air-intake fan. Unfortunately, the fan drew in air smelling of sewage.

Easy Ways to Upgrade Low-Voltage Systems

Industrial buildings constructed up into the 1950s were being built when energy was much less expensive than it has been in recent decades, not to mention its recent pricing. They have high ceilings, a characteristic that allows renovation designers to hide wiring in raised floors and dropped ceilings, which is useful when replacement seems a better alternative than repair, as is commonly the case.

Assorted Errors

After beginning with the around-the-corner mundane, this chapter proceeded to discuss some legitimate but highly unusual types of industrial equipment. It closes with certain types of installation encountered in commercial and industrial settings that are simply wrong and certainly not unique to old buildings. However, just as with residences, the longer a commercial or industrial building has been around, the more likely it is that violations have crept in without an inspector coming by.

Overly Isolated Grounding

One simply wrong design is the isolated ground circuit whose ground is overly isolated. NEC Section 250.6(D)—formerly 250-21—authorizes alteration or, at most, removal of some of equipment's grounding paths in the case of objectionable current. It does not, however, authorize leaving equipment ungrounded or bonded to only local grounds.

Bad Reactions to Noise

I don't know whether it still happens, but some manufacturers—or at least distributors—of electronic cash registers used to misunderstand this permission. They advocated bonding their equipment, and the circuits feeding it, to a grounding electrode that was completely isolated from that of the service or derived system. Consequently, since the 1990 edition of the NEC, in which CMP 5 accepted a proposal on this subject by Warren Lewis, Section 250-21d—now 250.6(D)—has specifically excluded currents that cause noise or data errors from being classified as "objectionable," Note that this is different from 250.146(D), which has authorized the use of Isolated Ground receptacles and the wiring feeding them (which most certainly are grounded to the system ground, if not a subpanel) precisely in response to concerns about noise. The restriction in 250.6(D) and its Informational Notes (fomerly *fpns*) means that I have no authorization to remove any part of the grounding electrode system because of stray currents passing through it. I can install isolating transformers on circuits, but they are permitted to isolate the power, not the ground. Any installation that was done differently is a candidate for correction; it does not warrant grandfathering.

The dangers associated with a completely separate grounding path are that the impedance between such a path and the main bonding jumper is too high to ensure operation of many protective devices and that it's too high to maintain a reasonably equal ground potential between the isolated equipment and nearby surfaces.

Ill-Considered Reactions to Pinholing

Another reason that installers have given for disconnecting grounding and bonding paths is fear that bonding the electrical and the metal water system could cause damage to plumbing systems. They are mystified as to what causes pinholes in copper piping, and their solution is to blame bonding and, consequently, create an electrical hazard. It's simply not so—except with DC power. Nonrectified AC can't cause ions to migrate in any particular direction, such as from copper tubing into water.

Too Much Tightening

I promised to return to overtorqued connections. On commercial jobs, with relatively high ampacities, there is some temptation to retorque connections. The best experts' advice on this issue is: "Don't." Thermography finds loose connections. Some lugs are given two torque values, one for new and one for retorquing. However, my colleague Ed Bihlear reports that these are primarily for use where a connection is reestablished after it has had to be taken apart. Still, several jurisdictions—including some in my area—require that switchboards, for example, be periodically retorqued.

Thoughtless Risk-Taking by Customers

Some other dangers that I point out to customers are found in offices—including plant offices—as frequently as in homes. They tend to be responses to insufficient wiring. Examples include reliance on extension cords, bypassed overcurrent devices—a short length of copper plumbing tubing probably is the most common substitute for a cartridge fuse with cylindrical ends—and overlamping. Occasionally there are special dangers. For many years, I serviced a chain store whose basement lighting has grown dim, because of its many burned-out fluorescents. I decided that I won't work down there this year, because the (storage) basement's been continuously flooded. They figured it didn't affect business.

Risks Created in Commercial Occupancies by Careless Electricians

As can happen in any occupancy, fuses and even circuit breakers have been simply upsized to avoid nuisance. It is equally common, though, for me to find branch circuits properly protected where they originate, but undersized starting at some point downstream.

The person performing a panel changeover may have lost track of a circuit's overcurrent protection and fallen back on being guided by the ampacity of the wiring at the panel; this is not good enough, though I admit I've made the mistake myself. At least I think it was a mistake, in retrospect; the circuit fed a couple of small motors and their markings were not clear.

Multiwire Misapplications

Article 605 forbids the use of multiwire circuits to power cord-and-plug-connected office partitions. It doesn't matter that they are permitted in other prewired partitions, even free-standing ones, as long as the circuits are set up for simultaneous disconnection. Confused by crowded panels and subpanels, unthinking installers working on commercial jobs are at least as prone as are electricians working on residential jobs to misapply multiwire circuits. Until adoption of the 2005 NEC, commercial installations did not generally require handle ties on breakers serving multiwire (as opposed to polyphase) circuits. Consequently, it is easy to find a neutral inadvertently shared between two or three energized conductors that turn out to draw power from the same leg or phase of the service or feeder. In the early days, but not in many, many decades, the solution would simply—and quite legally—have been to upsize the return conductor.

Breaker Misapplication

The next problem I encounter, more frequently in commercial establishments than in residential, is a particular type of inappropriate protection. This is likely to arise where someone upgraded service equipment, or at least loadcenters, or installed or replaced disconnects incorporating supplementary overcurrent protection. It is easier to detect and handle when installing new utilization equipment and is more likely to sneak up when just replacing a panel.

The mistake is protecting equipment improperly—usually with a CB when its instructions include restrictive wording such as, "Protect with a time-delay fuse rated at no more than 15 amperes."

It is devilishly hard for inspectors to catch this sort of error, when time is limited. I don't fault them strongly when I run into it. When I find this, I simply need to add fuses as supplementary overcurrent protection. This has been as simple as a fuseholder mounted in a handi-box—unless I also need to add a local, switched disconnect.

A Hawai'ian colleague, Mike Last, reported using a downstream fuse to protect a circuit simply because he did not have a suitable 15-amp circuit breaker on hand, just a 20. As all the 14 AWG wiring was downstream of the supplementary fustat, and the Listing instructions did not specify breaker protection, it was adequately protected.

> The breaker misapplication I run into far more commonly was mentioned in Chapter 6: installing a breaker that fits and conducts, but that is neither Listed nor Classified for the panel.

Anonymous Installers: Trouble They Create

The interiors of commercial, industrial, and institutional facilities most commonly have been repaired and upgraded over the years by maintenance personnel, people who generally never studied the NEC and who never have had to earn or maintain electrical licenses. Master electricians intimately familiar with major institutions in Washington, DC have mentioned finding this. Over many years, they had never seen an electrical permit pulled for any of the changes to the electrical systems of these institutions, including critical public institutions.

I certainly have looked over and worked on minor institutions. A public swimming pool allowed the installation of storage cabinets in the free working space in front of panelboards. I document this in Fig. 7-22. They provided GFCI protection for convenience outlets. However, one of the outlets in the changing room lost the protection of complete enclosure (Fig. 7-23), and its cover stayed broken for years. (I believe it was deenergized—I hope intentionally. Still, it should have been removed.) A church blocked panels and left their covers off till they were finally misplaced altogether, as seen in Fig. 7-24. A public county ice rink had a transformer located off the edge of the ice, to serve its refrigeration needs. The transformer blocked access to its own disconnect, shown in Fig. 7-25, and its location meant that free access to the transformer, in turn, was blocked by a column, shown in Fig. 7-26.

FIGURE 7-22 This public facility installed cabinets that block panelboard access.

FIGURE 7-23 Even if dead, a damaged GFCI should come off the wall.

FIGURE 7-24 Panel covers simply were missing.

Example: A Bad Sign   I was involved in the repairs required at a fifty-year-old church in Washington, DC. This discussion focuses on a few aspects of how the church's internally illuminated street sign was wired.

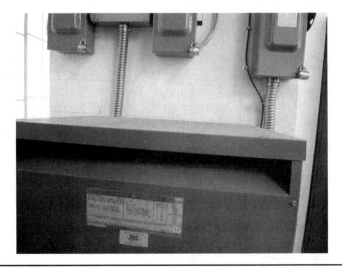

FIGURE 7-25   Lean over a transformer to work in its disconnects?

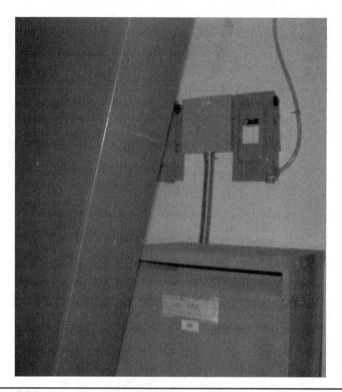

FIGURE 7-26 As installed, equipment's working space included the column.

Parenthetically, after fifty years, the rubber-insulated armored cable was badly deteriorated at outlets, even where there was no reason to assume circuit overloading, or the nearness of heat-producing equipment, had contributed to insulation degradation. Fifty years is pushing the limit, or, truthfully, beyond the limit, for most rubber.

It would be reasonable to expect the body of such a sign to constitute a moisture-proof enclosure, rated NEMA 3R or better. In practice, this one fell short. Perhaps some part of it had failed over the years; more likely, the way the angle iron had been brought into the top of the sign when first installed had provided an ample path for rain, whenever there was a driving wind. At least this was what I tentatively concluded. The interior showed clear evidence of moisture and, indeed, the terminals for its two 6-ft. fluorescent lamps had corroded.

The installer had wired the sign with lead-sheathed armored cable, run out through the wall of the building, along the supporting angle iron, and into a conduit body at the bottom of the sign.

Some Concerns
- Conduit bodies are not a generally approved means for making a transition from armored cable to other wiring methods.

Figure 7-27   Were there two violations or three?

I'm not saying that AHJs should forbid the practice. However, Section 314.16(C)(2) forbids enclosing splices in a condulet, except in the rare case where it is marked by the manufacturer with its volume. Before this was permitted in the 1978 NEC, splices were forbidden in LLs, LBs, and C-condulets. This may not be a problem here, because back when the sign was originally installed these prohibitions had not made it into the NEC! Even now the splicing aspect would be okay if the transition was made by stripping enough of the armor back to reach the cable's conductors through the condulet and into the sign before being spliced, but the requirement that conduit bodies be rigidly supported may well be. In this case, it's not stripped (Fig. 7-27), so the conductors aren't protected at the cable's end, inside the sign (and the armor has to make a somewhat sharp bend at the end of the condulet).

- While ACL ("BXL") cable installation was permitted in damp locations, and for use where exposed to weather, there was no evidence that the opening through which it left the building ever was sealed.

- It entered through a standard BX connector. The setscrew connectors are just for indoor use. When connectors are Listed as "Dry Location Only," it's no sin to use them outdoors, but only in locations that are sufficiently protected to keep the connectors dry enough that they might as well be indoors.

- The conduit body was secured to the sign not with a chase nipple but with another BX connector passing through from inside the sign. BX connectors are not Listed as substitutes for chase nipples.

To complete the story of the sign's woes, there were a few more-serious problems:

- Its ballast appeared to have cooked;
- One of the wires in its fixture channels was badly charred;
- In violation of NEC Sections 110.3(B) and 110.27(A)(2), formerly 110-17(a)(2), the controlling time clock was missing the protective interior shield that guards the energized terminals from contact when you swing open the front cover to reset the clock, or to turn the load ON or OFF manually. Someone working on it either lost track of that part or decided that it was an unnecessary extra. Since the clock was working at the time, apparently they presumed that any leftover bits could be dispensed with;
- Finally, and serious only from the perspective of functionality, the time clock was not working.

How did the sign come to be installed in so questionable a manner, right on the front of a stately church? A reasonable suspicion is that this came to be because the sign is a good 20 ft. in the air, and the inspector was not eager to clamber up there, even assuming that a ladder was at hand when he came by. Besides, the sign may have been installed after the church was complete. Sign installers are not the most by-the-book trade when it comes to pulling electrical permits and calling for inspection.

Here is what was done to remedy the situation. First, I had to decide how to get the sign working. While I could have at least replaced the ballast and lamp terminals, it seemed wiser to replace the entire weather-worn fixture, containing one ballast and two 72-W lamps and lampholders, with something more standard. There was ample room for two high-output, 4-ft., two-lamp fixtures with 0° ballasts.

These were not outdoor luminaires, but then the originals had not been, either. As originally designed, the enclosure apparently was considered adequate to keep them dry. Recall that I had determined that most likely moisture had entered the sign at the top, where a hole had been cut to bolt the supporting angle iron; I caulked that hole. I then removed the ACL running outside the building, and (although not in this sequence) terminated it in the back of a bell box I installed right on the outside wall. I wired the sign using UF cable, run from that box. On the bell box, I mounted a photocontrol, Listed for the use, with a rating well over the wattage of the ballast installed in the sign.

Inside the church, I bypassed the dead time clock and mounted an enclosed switch to serve as the disconnect. I pulled back the ACL, cutting back the sheath to where I was delighted to find perfectly good conductor insulation. I fed these conductors into the bell box through a BX connector. After tightening its setscrew against the armor, I pulled back on the cable so as to recess the connector into the wall, as best I could, before securing the box. I then caulked the devil out of the space behind and around the bell box, so moisture could not weasel its way either into the building or into the connector, and thence into the box.

Other options would have been to use conduit, EMT, or flex. Sealtite would have been a fine choice, but given that the KO was on the underside of the sign enclosure, even regular flexible conduit would have met *Code*, so long as it enclosed THWN. However, it would have rusted away fairly soon.

Changing Rules: Neon

A different sort of problem to watch for derives from changing *Code* and Listing requirements.

Neon-type lights are bent, and installed, by specialists, creating Listed units. Notwithstanding the fact that they ignore end-users' clear intentions to violate the NEC, many of these folks tend to know what equipment is suitable for use with their installations. They are neither more nor less likely to ignore Listing requirements than is an HVAC installer. Figure 7-28 shows a result of electrical inspectors not

FIGURE 7-28 I still don't get why this kluge worked.

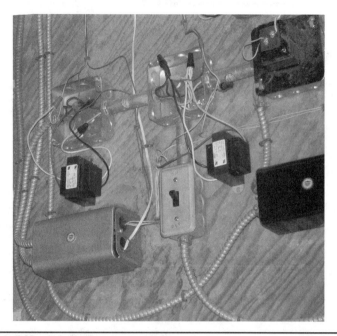

Figure 7-29 At least there's grounding continuity.

checking HVAC installations. Upgrades are even less likely to get looked at, resulting in careless violations joined to good old work, as in Fig. 7-29. Not to leave out another sometimes—deserving trade, Fig. 7-30 shows a plumbing installation (Fig. 7-31 is a closer view of the otherwise-legal disconnect).

Even though neon installers may have known what they were doing, I don't dare assume that old outline lighting fixtures comply with modern standards. Around the time that the 1999 NEC was adopted, a new UL standard for neon transformers and neon power supplies began to be enforced. Standard 2161 requires that most equipment incorporate secondary ground fault protection. Even more to the point, NEC Section 600.23(B) requires this protection more often than not.

This change has two consequences. First, I'm not a sign expert and, if called in to replace a burned-out transformer, I may not order the identical unit even if available. Also, I know that I don't dare take any chance that I'm working on it energized.

It's Been Worth Taking the Trouble to Find Out What the Customer Wants—and Needs

Example This demonstrates a general issue that applies to commercial jobs as much as to any others. A customer calls on me to perform a

Figure 7-30   Here's a blocked disconnect, seen from the right.

Figure 7-31 This disconnect was inaccessible from any approach.

simple change, a straightforward, make this place a-little safer matter of convenience, and is quite unprepared for its implications. When those implications mean too much expense, it can put the kibosh on the whole project. When the implications are more a matter of requiring significant planning and figuring, I've had a better chance of selling the customer on proceeding with the effort.

The owner of a high-tech service business asked what it would take to equip his newly acquired warehouse with power to the (fixed) workbenches. He and his partner probably would be the only employees, but they would be servicing a fair number of items at any one time: mostly consumer electronic equipment. I proposed running cable under the workbenches, popping it up to feed receptacles. The customer suggested surface metal raceway, by which he actually meant a multioutlet assembly. This made sense.

I told him that I would be glad to do the job and that pricing should be straightforward once the customer knew how long he needed his runs and how many circuits were required—presuming his existing panelboard had sufficient capacity. The number of circuits, I told the customer, legally depended on how many items might be worked on simultaneously, or might be in the process of burning in at the same time.

You may wonder why I was making life so difficult, rather than pulling a guesstimate out of my pocket. For one thing, the customer had ballparked the length of multioutlet assembly he wanted at 75–100 ft. Twenty-five feet is a considerable difference. More important, though, is the *Code* section permitting more-generous load calculations with receptacles that are part of multioutlet assemblies [NEC Section 220.14(H)—formerly 220-3(c) Exception 1]. Without some careful figuring, it could challenge the amount of space in his panelboard, if not realistically its capacity. It can jolt a customer to learn that one added load means the panel, if not the entire electric service, must be upgraded, or at the very least a subpanel must be added.

I looked at the worst-case scenario: 100 ft. at 180 VA per ft. gives 18 kVA. Divide this by 2400 (120 VX 20 amps) and I get 7.5–meaning 8– circuits. This would call for two 20-amp circuits serving every $12\frac{1}{2}$ ft. of work space. This seems like considerable overkill for two men's work, unless they would be simultaneously, continuously, powering equipment that has a significant number of heating elements or of fair-sized motors—or unless they figure there is indeed a chance some item will blow up periodically, and it would be nice for the whole lot not to lose power when this happened. (Supplementary fuses could help here.)

I also considered the other extreme. I could run one circuit in each direction from the panelboard and assume that each man will work on only one piece of equipment at a time. (It just is not credible to propose that only one partner would be working at any one time.)

Now Section 220.14(H)(1) permits each 5 ft. of multioutlet assembly, rather than each 1 ft. to count as a 180-VA load. Every foot is a 36-VA load, if I ignore fractions shorter than 5 ft. Now I can feed 66 ft. of multioutlet assembly in each direction.

Unfortunately, it is hard to believe this scenario as well. Sometimes, each man will be working on a piece of equipment and each will have two other items plugged in and drawing power at once. These might be a 120-volt tester and a power tool; or a couple of modules being burned in. Even if the receptacles in the assembly are duplex, which, in other cases, allows the two receptacles sharing a yoke to be counted as one outlet, I dismiss that. There is little reason to assume that the CMP meant for the definition (multiple receptacle) to apply to 220.14(H) concerning multioutlet assemblies. Does this mean I should return to the first, more stringent calculation? Not necessarily.

There are two ways to avoid that; either would require consulting further with the customer. Multioutlet assembly can be purchased with outlets spaced more than 1 ft. apart. Then, instead of applying the exception, the general rule in NEC Section 220.14(I)—formerly 220-3(c)(7)—of calculating 180 VA per receptacle outlet can be applied. This would bring the spacing up to 20 ft. of multioutlet assembly per 20-amp circuit. This might well work, although it would not be a useful option where the customer wants an outlet every 6 in.

A second possibility would be to ask, "Is there any section of workspace where, in the ordinary course of events, you expect to just store equipment rather than work on it?" The answer might be, "Can't say." It might be, "No, I meant 75–100 ft. of that molding with outlets spaced all the way throughout it." However, if the answer is "Yes," the less-used section can either be run in surface metal raceway, perhaps with an outlet or two, or run in a multioutlet assembly with its load calculated at 180 VA per 5 ft.

There are a couple of other possibilities. One is the option of cutting the number of home runs, in the event that eight circuits are required, by using multiwire circuits. Multioutlet assemblies do come in a two-circuit, multiwire configuration. In this case, though, a judicious electrician would be reluctant to employ multiwire circuits. This would be due to the risk of the electronic equipment's power supplies' overloading the neutral and also the risk of sending noise through the neutral from one piece of equipment to another.

Anticipating Change

A final consideration is this: if I calculate and install based on a two-person shop, what happens when the owners are surprisingly successful and hire more help? This could mean more intensive use of the workspace and much heavier electrical loads. Sometimes saving the customer money up-front is no boon. Other times, it makes sense

FIGURE 7-32 The bollard was too far for protection.

because when they are ready to expand they will have the capital to pay for additional work. Where everything is being installed in surface wiring, as in the job just discussed, coming back to do more work does not create the disruption that it would if changing the configuration meant that walls had to be ripped open.

It is critical not to just give lip service to requirements. Figure 7-32 shows a service raceway in a driveway, using EMT, but protected by a bollard at each end and well supported. Figure 7-33 shows that the

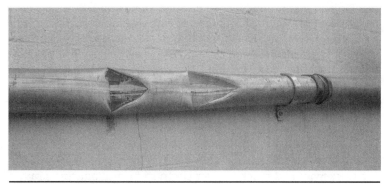

FIGURE 7-33 Here's the damage in close-up.

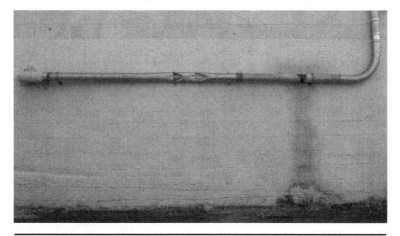

FIGURE 7-34 Trucks dinged this.

bollards were too far to protect the EMT from damage; Fig. 7-34 is an overview of the run.

Example Here's a different sort of case, where I was asked what it would take for me to come in as a consultant, perhaps to be deposed at a forthcoming civil trial.

The caller owned and operated three adjoining businesses, a garage, an auto parts distributor, and a body-and-paint shop, each separately metered. However, the outside lights on all three parts of the building were fed from the garage's panel, through the same controller. At one point, he had leased the three properties to people who wanted to run these businesses themselves. The leases forbade alterations.

Sometime later, having reacquired two of the businesses and, I suspect deep inside, wanting to reacquire the third (the parts store) as well, he called me. He wanted an investigation and report on the legality and safety of alterations performed in the third business, performed without permission and, he suspected, unsafely—and naturally without permits or inspection.

He pointed to one: the lights outside the parts distributor seemed to be separately controlled. At any rate, they did not extinguish with the others when he operated the master controller.

There were various problems with what he wanted, such as his insistence that one-half hour should be enough to evaluate the parts store's wiring and panel, leaving me one-half hour to determine what permits had or had not been pulled or remained open, and then type up a report of all the findings, suitable for use in court. I brought this up, though, to point out two issues. First, while controlling the lights from one panel might have been legal when he was running all three businesses and paying the bills, once he sold the businesses and others were paying the bills, a violation was created by having control of the lights on one part of the building in the hands of a business in the other part of the building, originating from a different service. Second, businesses reconfigure their layouts and wiring every so often, especially when new owners or management take charge. All too often, there is no professional oversight over the work and no paper trail.

CHAPTER 8

Accept, Adapt, or Uproot?

This chapter talks about distinguishing what the customer may optionally choose to change from what needs to be changed. I examine two problems: being asked by a customer to make changes that will bring a system into violation; and dealing with crowded enclosures. I begin by looking at a challenge specific to the installation of fans and heavy chandeliers: retrofitting support required by the NEC. Other chapters explore the general issues raised in this chapter from different angles.

Paddle Fans and Other Upgrades

The following discussion focuses on paddle fans, but much of it also applies to other utilization equipment and outlets as well.

Suppose that my customer says, "I'd like you to hang this fan from the ceiling in my (old) building." Until the *Code* changed, whenever installing a paddle fan, I would perform a pull-up on the fan or fan support and make sure that nothing gave way. The difference in weight between even a relatively slim electrician and a fan compensated for the fact that a dangling installer is not the same as a long-term dynamic load. The Veterans' Administration used to rely on a specification that fan supports be sufficient to hold up a 150-pound static load.

Boxes

The old approaches no longer are necessary. The instructions that come with many paddle fans specify that they are only to be supported from boxes Listed for the purpose. No inspector has given me grief for using Listed fan supports per spec, or for mounting a fan directly to structural elements, through or past the enclosure.

I find it far more common in older structures than in newer buildings for a ceiling outlet to be mounted directly under a joist.

Figure 8-1 This is a typical amateur smoke alarm retrofit.

Unfortunately, that ceiling outlet often has an undersized enclosure, one that I have to consider replacing, or no enclosure at all!

Crowded Boxes

Here is where a whole new issue is broached: volume. Pancake boxes or ceiling pans (I've found them at wall sconces too) come in two sizes: 1/2 in or 5/8 in deep by either 4 in or 31/4 in in diameter. (They used to be made in smaller diameters as well.) This means that, at best, their volume allows only two 14 AWG conductors. There can be no devices, no internal cable clamps, no hickeys, no incoming ground wire. Conscientious electricians may have only used them with canopied fixtures, even before the days of the occasional volume-marked canopy. Ignorant people usually get it almost right, at best, as they did in Fig. 8-1.

I have replaced a light mounted over a pancake box without replacing the box, although many pancake boxes, because they contain very old rubber-insulated conductors, are worth pulling out just so that I can get to good insulation. Figure 8-2 shows badly deteriorated conductors that had been connected to the leads of a modern

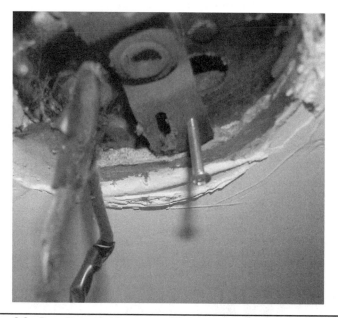

FIGURE 8-2 Decaying rag wiring serves 90-degree C luminaires.

luminaire that specified insulation having a 90°C temperature rating. Figure 8-3 shows what I sometimes do to strip back very short BX—too short for sawing—in trying to reach insulation that's not badly deteriorated. Unless I damage something in the process, replacing an undersized box with a modern, deep one makes a better job. Even so, conductor temperature ratings limit the choice of replacement luminaires that can legally and safely be installed directly over the old wiring.

How about a ceiling box, 3 in or 3¹/₄ in round or octagonal by 1¹/₂ in deep? No longer made, its volume is listed in Chapter 11, Grandfathering, Dating, and Historic Buildings. Very commonly it will have been used as a junction box. Say it has one cable coming in, carrying power onward, one going down to a switch, cable clamps, and a fixture hickey (the hickey is attached to the hanger bar supporting the box between ceiling joists). Is the box illegally overcrowded? Probably, but I won't be too hasty to judge. It depends.

Assume that the conductors are 14 AWG and the cables are old BX: there are no grounding conductors—or bonding strips, for that matter—in the cables. Every item—each of the six wires, the pair of clamps, and the hickey—demands 2 in³. That comes to 16 in.³ all told. The maximum volume the box can possibly offer is about 14.5 in³, but once the *Code* specified fill volumes, it allowed less than 11 in³ for these. I rarely see volume-marked fixture canopies, but if using a roomy, but

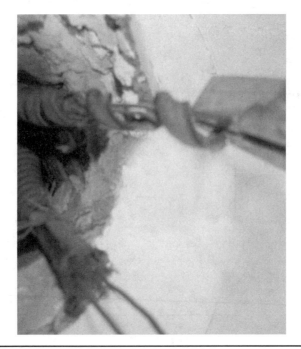

FIGURE 8-3   I strip very short armor by unwinding it.

unmarked canopy, makes the job acceptable to the inspector, that's reasonable in my view. I may buy it, on an inspection or consultation. I just don't bother replacing an undersized box with a close-to-adequate box, though, whether or not there is a canopy. On occasion, though, I will decrease crowding by redoing one of the excellent, but bulky, old splices, as in Figs. 8-4 and 8-5, where I found a fat splice and unwrapped the friction tape so I can crumble away the old rubber, then trim back the twist and clean it up to attach a modern wirenut.

I can't be sure of passing inspection with a crowded enclosure. Sometimes I'll add volume to a ceiling box or wall outlets, for that matter. If the box is sufficiently, illegally, recessed that I need to add a box extension to bring it out to where it will be flush with the surface, the extension can double the available volume. True, the extension must match the box; it has to line up well enough to provide complete enclosure. On rare occasion, I can move the box back up into the ceiling so that I can add a box extension. On other rare occasions, generally in walls rather than ceilings, I can add Wiremold®-type extensions. Or, with a customer's okay, I can let the extension jut out a little. Figure 8-6 shows one of these. In such cases, goof plates can come in handy. The extension in Fig. 8-6 solved a second problem. As Fig. 8-7 shows, the original box not only was recessed but had been jammed so its ends

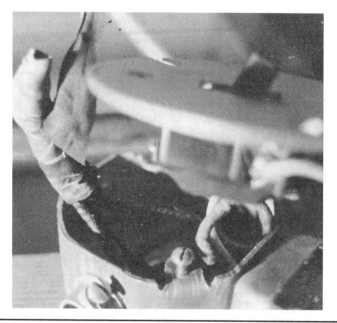

Figure 8-4   I can make room by redoing big old splices.

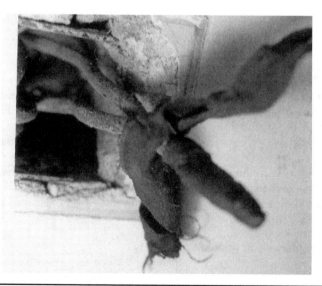

Figure 8-5   I start by unwinding friction tape.

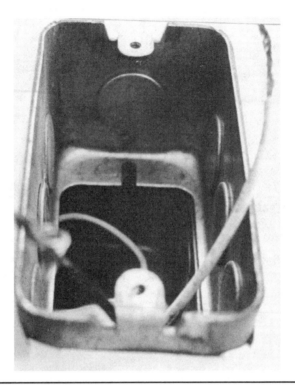

Figure 8-6   Sometimes a utility extension box fits.

were bent. As a result, one ear was no longer parallel to the other, and also its 6-32 holes no longer were quite far enough apart for the screws holding a device yoke to reach.

Often, in a ceiling, a fixture canopy will cover the protruding bit. The instructions accompanying some paddle fans explicitly instruct the installer to leave a small gap between canopy and ceiling. I have confirmed with UL that such instructions are part of those fixtures' Listing. In a wall, a goof plate may do it. Some odd ones are available, such as shown in Fig. 8-8. Replacing the box is ultimately cleaner, but a little riskier to undertake, often messier, and almost always more expensive. It frequently has an advantage in that I am less likely to overlook existing damage to conductor insulation. This is coupled with the disadvantage that the handling may overstress fragile conductors or insulation.

Box Replacement: Additional Benefits

With some boxes, replacement is the only reasonable choice. Even when the case is not quite that extreme, replacement often is worth considering. Here is the great advantage of replacing a box. Normally,

FIGURE 8-7
Panelers can
squash down ears
and recess boxes.

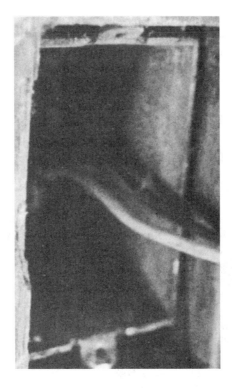

FIGURE 8-8   A goof plate or a deep cover may be acceptable.

I am replacing the overcrowded enclosure with a wider box— sometimes even an 11B—not merely a deeper one of the same diameter or even shape. The increase in diameter or width means that even when wires coming in from a side are way too short, even if they do not look good, meaning that the insulation appears shot, I've given myself some leeway to remedy those problems. I can cut back the cables at least $1/2$ to 1 in to gain access to more wire and to access insulation that has not suffered the same stress. If necessary, I may have a chance to free up at least another $1/2$ in per conductor by changing over from internal clamps to cable connectors. Even better, having pulled the old box out of the ceiling, very frequently I can loosen a staple or two enough so that, with care, I can safely tease a couple of more inches of cable slack into the box.

Replacing the box with a bigger one can make quite a difference. Doing so can mean changing from a box containing crowded splices of rotten conductors 2 in long to a roomy, easy-to-work-with box, manufactured with a tapped 10-32 grounding hole—a box that contains several inches of reasonably good conductor. I talk elsewhere about temporarily fixing conductors—which usually means taping. If there just is not that much conductor to be had, at least the new box gives me room to pigtail lengths of healthy conductor onto the tired old wires. When I need to heal the existing conductors, cutting back the cable sheath can expose fairly good insulation to grab with the end of my tape (or shrink-tubing).

If the box is fed by conduit, I may be somewhat less inclined to replace it, especially if multiple conduit runs enter the box from opposite sides. How do I get it out and, furthermore, fit a bigger one in? Sometimes I can muscle them. Besides, with conduit, I can pull out shot wiring and pull in new conductors, long enough to leave 6 in or more of free conductor at each end, and with better and more compact insulation.

Replacement: Problems of Chopping into Joists   Suppose I want to replace a pancake box, mounted right under a joist, with a larger enclosure. What I install almost certainly will be at least $1\frac{1}{2}$ in deep. (For a considerable premium, for one job, I did get hold of $1\frac{1}{4}$ in deep 1900s, to use with flush or shallow mud rings; $1\frac{1}{4}$ in deep 1900 extensions are also available.) This may mean I have to cut into the joist, unless the existing box is deeply recessed because of bad workmanship. More commonly, I find a box recessed because a layer or two of ceiling was added to the original surface. A deeper box can use that depth.

While I may need to chop a bit, I can't weaken the structure. That can happen: the underside of a joist serves as the primary tensile member, keeping loads above it from breaking it.

I will cut even an inch out of the bottom of an old 2×10 or 2×12—though not a modern manufactured joist constructed of 2×2s or 2 × 3s, top and bottom. I've never found one in an old ceiling, not even sistered to bad wood. Engineered wood products were first used in the 1950s, made of solid, glue-laminated boards. I've never run into one in an old building, but I wouldn't cut or notch anything suspicious.

One alternative I've used is fan support boxes that straddle a joist. The last time I did this, I actually had to install a new box away from the joist, serving as a junction box (j-box), with a blank cover. A few feet of NM to the box straddling one side of the joist and I hung the fan. I didn't like the j-box being above fan blades, but it was accessible. This was confirmed, ironically, when I had to go back to it, troubleshooting the circuit.

Inspection note

While the homeowner might like the idea, a throwaway such as the ones I found, shown in Figs. 8-9 or 8-10 is never acceptable. I don't even think the j-boxes shown in Fig. 8-11 passed an inspection—in 1962. I located these once I removed the range top, which was held in place by tightening one setscrew under each removable burner. However, as Fig 8-12 shows, the j-boxes were so far down, nestled against the side of the recessed oven, that they only could be reached by undoing the wood screws mounting the oven front to the cabinet, and hauling the oven out onto the linoleum. (There were other violations associated with their installation, but those are shown in Chapter 6.)

FIGURE 8-9 Out-of-the-way j-boxes need finding and fixing.

FIGURE 8-10
Throwaway j-boxes
are common.

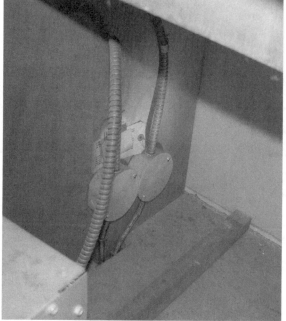

FIGURE 8-11 Junction box locations can surprise me.

<small>**FIGURE 8-12**</small>   Behind/alongside an oven and under a stovetop was one.

Replacement: Other Issues   When replacing an undersized ceiling box, perhaps because of bad insulation, I have used a number of options. For instance, in one case the ceiling was open enough, and the joist was so far recessed, that I sistered a couple of short lengths of 2 × 6 against the side of the old joist, bringing the structure down just far enough that I could mount my fan-support box very securely against this new structural element.

Here is a problem I've run into when attaching a new box, or other equipment, to old structural elements. Compared to newer wood, old wood is brittle. This is true of wood lath, as well as joists. Therefore, I find I'm far better off sawing than whacking, chiseling, and gouging. Sure, I've worried away the occasional 1/2 in of joist or stud that interfered with my setting a box flush, using the tools that were handy. Sometimes this hasn't worked too well. It has cracked even when I first slit the joist on either side of the section to be removed, using my Sawzall™! On rare occasion, I've needed to drill pilot holes before screwing into dried-out wood.

Less-Than-Ideal Locations

Sometimes the customer wants a paddle fan and the existing outlet is not at a joist. In a number of situations, albeit relatively-rare ones, I don't have the option of replacing the box with one Listed for fan support.

The most common problem with paddle fan boxes and old wiring is that, if the existing box is very crowded, the available fan support boxes may not have enough volume. That's when the j-box comes in. There always will be cases where I have to warn the customer that the ceiling will need to be chopped open and replastered.

Another solution to the problem of finding a fan support that will accommodate existing cables depends on how flexible the customer happens to be. When working below an unfinished attic, on rare occasions I have been able to safely rotate the existing box up so it remained accessible from above, rather than below, the ceiling. I've also used this approach when converting to a luminaire with post-1987 temperature requirements. This way, the junction box did not require a blank cover visible from below. A bonus is that I had room to add an extension box, out of the way.

If my location is right between joists, under an unfinished attic, it's been easy to to nail a 2 × 4 between joists just where it's needed. Otherwise, when I'm not faced with the need to accommodate more than one or two cables, an aftermarket support bar kit usually works. I've found several limitations:

- I've proved vividly to myself that I do have to set it in straight; otherwise, tightening has just made it ride further away from perpendicular, as its teeth wouldn't bite the joist. I've run into old joists that were closer than 13 in from each other; not enough room for the bar between them.

- I've noted that drywall backer board, and sand mix, have keys that push the bar back far enough that my fan support box ended up recessed too far.

- Spreaders (joist braces) can get in the way; I get rid of them.

- A strap, once having supported the old box, can block the bar; I have had to remove this, too.

This is a good place to examine the ideas of solid support and solid contact. It took the NEC many years to get around to saying that one 6-by-32 screw in the center of a receptacle does not provide an adequate path for bonding it to a surface-mounted box. Contact with a flush box used to be considered adequate to bond devices, although this has come in and out of the *Code*; only with the 2005 *Code* was the "metal-to-metal" aspect refined.

For fan support, I'm uncomfortable using anything less than, minimally, #8 or #10 screws, penetrating at least $1\frac{1}{2}$ in into solid wood. When I have any doubts about of head size versus mounting hole size, I use a washer under the head of a mounting screw—especially when, in some odd circumstance, I use flat- or panhead screws.

Old Boxes and Their Accessories

Two more issues that have not yet been addressed properly come up frequently in dealing with older boxes. One issue is the need to accommodate dated equipment that was not designed to mate with currently used products. The second is the tension between two options.

One is merely doing what will meet *Code*, and thus minimizing present mess and expense; the other is going further to do a better job.

Boxes Lacking Integral Provision for Covers   The first, and more-contained, issue is dealing with antiquated enclosures. One piece of equipment that mates with nothing modern is a box without screw holes at the perimeter or ends or corners. It has no ears to which to attach devices, rings, extensions, or covers. Some old pancake boxes, as well as some deeper round boxes, were designed this way at a time when fixtures were mounted on hickeys rather than straps. Occasionally, I replace an old fixture that was mounted to one of those. Without replacing the box itself, how am I to attach fixture canopies or straps?

I've used the hickey to securely support a strap that incorporates tapped 8-by-32 holes. Even where the fixture ends up more than 7 ft. AFF, I'm not crazy about the fact that the fixture relies on the support for grounding contact. I have drilled and tapped the box and used mounting screws long enough to reach to those holes in the back of the box. The equipment may end up a little off-center, but as long as the enclosure is covered and wiring is not threatened, it works. If there's a way to bring a bonding wire to the box, I'm happy with the hickey as support. If necessary, I've drilled and tapped one 10-by-32 hole.

By far the poorest choice, especially in terms of safety grounding, is to take wood screws directly from the cover into the building structure, in back of the box. When I find this, I won't continue to use it if I am doing any work at the outlet.

Older Style 4 in Square Boxes   Another misfit piece of equipment is the very old 4 in square box, whose tapped ears are just not quite where they need to be to line up with a modern device ring, plaster ring, or extension box. There are a number of reasons I may mate modern equipment with such an old box.

- I could be adding or removing a device and thus need to go from single-gang to two-gang or vice versa.
- I might be adding a cable or raceway, and thus require the additional volume that would be afforded by adding a box extension.
- I could notice that the box and ring are recessed beyond the legally permissible depth, and want to set it right with a box extension or a deeper ring. This problem is ever so common.
- Finally, I may have an old box with an old mud ring, one that, like some old pancakes, is designed without tapped ears, and want to replace the existing fixture with one that does require a strap or ring with tapped holes.

Since I have been doing old work for a while, I have accumulated a collection of old device rings that will solve some of these problems. (I've even held on to some rings that convert 8Bs' openings to 3½ in or 3¼ in, which I've kept primarily for their ability to add ½ in. or so of depth.) Inspecting, I would see no harm at all in modern device rings drilled and tapped so as to place the holes where needed—or very close to where it needs to be. An eighth of an inch mismatch would not concern me—but a ¼ in miss probably would not line up with a reasonable amount of "persuasion." Technically, such a modification might be a Listing violation, but I'd call it a reasonable accommodation. I won't attach a modern ring or extension to an old box catty-corner, or accept this on inspection.

Another type of box that does not match what normally is used currently by professionals—although I know it is available—is the round or octagonal box that measures 3¼ in in diameter by 5/8 in deep. True, modern mud rings still have a 3¼ in opening, as do some fixture canopies.

Extension Boxes   This is a good place to discuss extension boxes in greater detail. Extensions certainly do add wiring volume and, in the rare case where a box is recessed 1½ in or more behind the surface, an extension can do a nice job of bringing the front of the enclosure forward. On the other hand, I don't consider using an extension box quite as good a job electrically as installing a box of the right size at the right location. I'm primarily concerned about access to wires and splices and also about the grounding path (unless there's a ground wire to the back). I can imagine a case in which I might reject a job where a 1½ in extension box had been added and I could not pull out a pigtail attached to a wire or a multiwire splice for testing—ideally 3 in or more.

Mating Fixtures to Boxes and Protecting Combustibles

Another problem with a small wiring enclosure, such as a 3 in round box, is that a luminaire's wiring compartment, which includes everything under the canopy, extends to the sides, beyond the box. When the surrounding, included surface is combustible I simply don't find it protected with something noncombustible. If I'm doing the installation, I *carefully* add some sheet metal.

New Fixtures Meet Old Wiring

When it comes to replacing existing fixtures, paddle fans are far from the only problem children. Whenever a customer calls and says, "I have some fixtures I'd like you to put in to replace what I've got," I respond with a number of questions before making an appointment. Just as with a paddle fan, the first is, "Is it a wet or damp location?" I explain that screened porches normally are not considered dry locations. Here's another critical question I ask: "Look at the instructions that came with the fixture. Does it say anything like, 'Use only with supply conductors rated for 90°C or 75°C'? If so, how old is the house's wiring? Look everywhere for this warning, not just in the paperwork but on the fixture near where you put in the light bulbs, and even on the box flaps." Nobody will be happy if I show up to do the job and find that without rewiring, I cannot use the fixtures the homeowner planned on. I've had many customers return fixtures.

Old Does Not Mean Bad

Boxes are only one example of old designs that are difficult to match. Old does not automatically mean bad, even though antiquated designs can cause problems. When I come in on a consultation or a final inspection after a semi-gut, and find rewiring work such as that in Fig. 8-13, supposedly ready to be reenergized, I wish they had paid for someone with more knowledge to do a less-rushed job.

Occasionally, old is even better than new. Figure 8-14 shows a 30-amp SFR service still in use at the end of 2009. It had been installed with care: a weather-tight connector plus a drip loop, which is rare today. How many buildings today are wired in rigid conduit? Yet some old single-family residences get their power through conductors run in black iron pipe—which is just as good as galvanized rigid pipe, except for its inadequate rust resistance in damp locations. Plumbing pipe is quite a different matter, particularly when coupled and connected with plumbing fittings. Although the first

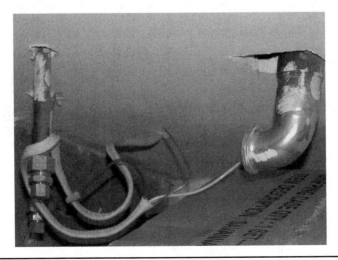

Figure 8-13 A look under a sink can say "not ready for Final."

FIGURE 8-14   Old, tiny services can show classy work.

electrical conduit was basically deburred plumbing pipe, it has since been illegal to use plumbers' fittings for electrical work. Long ago, existing gas pipes were converted to electrical use and passed any inspections that took place. Nonetheless, while what's there may be grandfathered, I would not pull new conductors past tight-bending plumbing fittings. I wouldn't feel comfortable passing such replacement work, either, unless it had been meggered, at the least.

Similarly, small services are not automatically bad. Not that many years ago, many modern, all-electric houses with heat pumps and electric cooking—fairly large middle-class homes—drew no more than 40–50 full-load amps (FLA). The fully-packed three-phase, 208/120-volt service in a small grocery served about half a dozen roof-mounted refrigeration compressors, as well as assorted other loads, without pulling more than 50–60 amps. (Since the time I wrote that, even this small shop added yet more refrigeration equipment, so I no longer believe that figure remains quite accurate.)

Certainly there is nothing inherently wrong with fuseboxes as circuit protection—although I have known insurance companies to require their replacement with circuit breaker services. Perhaps they've never seen a burnt CB like Fig. 8-15. When such a changeover has been done incompetently, the building is that much more dangerous.

FIGURE 8-15 Most damaged CBs aren't this burnt.

Besides, fuses don't decalibrate the way circuit breakers do. Still, fuse-boxes don't survive forever. In addition to looking and sniffing for evidence of scorching, on the advice of Littelfuse's Kenneth Cybart, I check for cracks or breaks in cases, ferrules, or insulators. He also recommended checking that fuse clips provide adequate contact with fuses. How? Ideal's testing expert, Brian Blanchette, explained that the way to check fuseholders for adequate contact pressure is thermal imaging while the circuits are carrying loads.

Having said all this, I still agree that, aside from maintaining calibration and having better intrinsic AIC capacity, an older fusebox is behind the times. This is not to say that fuses are outdated. In late-2009, Bussman offered a modern, switch-controlled, indicating cartridge-fuse panelboard for commercial installations. Using it still would require installing any AFCIs at some point downstream and doing without a common-trip option for multipole circuits.

Some Outdated Designs Do Have to Go

Unfortunately, there are some old designs that would be perfectly okay except for the fatal fact that they are out of date. If parts were available, this equipment could be repaired, rebuilt, and sustained

Figure 8-16   Never-used, not-very-old CBs can be out-of-date.

indefinitely—if it were sufficiently efficient and safe. Some pieces, however, are outdated despite having been recently manufactured and never installed. One example is the pre-2003 GFCIs in Fig. 8-16, which I never installed, and now wouldn't try to install to fill the GFCI requirement. Others are just are not worth keeping energized unless my customer loves electrical antiques and, in some cases, recognizes the risk. I don't doubt that they are available online, if not otherwise, and I suspect that unwary shoppers have been stuck with them, perhaps even installed them because they did not recognize the key flaw of being out of date. The same is true of the receptacle GFCIs in Fig. 8-17. Garage sales and flea markets may be sources of impressive discounts on electrical supplies, but saying this is not the same as saying that they necessarily offer products worth buying. The boast on the GFCI package, in Fig. 8-18, may have been true at the time of manufacture, but it doesn't stay true forever. Besides, I have found off-brand GFCIs to be too unreliable to install.

What does this mean for installations that contain outdated GFCIs such as shown Fig. 8-19, a picture of GFCIs that I certainly would not install today? I push the TEST button. If there were a concern about particularly high fault currents, perhaps because the utility had upgraded the transformer or because the risk of lightning strikes had been reassessed, I would recommend replacement. I would act

FIGURE 8-17   GFCI devices can also be new and outdated.

differently if I were dealing with a customer who was interested in having the latest and the best.

I do not globally include fuseboxes in the category of outdated antiques, except in that they don't allow for AFCIs. But soon I expect receptacle-type AFCIs to become available, like receptacle-type GFCIs, if the 2011 *Code* accepts a change that has been proposed. They have been Listed for many years, although not manufactured, because there did not appear to be sufficient demand.

The following question almost always is worth putting to my customer when the situation arises: "Do you want me to fix this or to replace it, given that, most likely, it will take as much in time and materials either way?" I need to think through the probabilities before proceeding, or even asking. If replacement may well be the less

FIGURE 8-18   It doesn't matter that they say they're up-to-date.

Figure 8-19 Outdated GFCIs that are in place I usually test and leave.

expensive option, I have come to realize, it is especially important to ask, before proceeding with the repair I am called in to perform.

Superannuated Devices

Here are a few designs that, while fine in their time, eventually ought to be replaced.

Despard Devices

Despard, or interchangeable, devices, Fig. 8-20, while still made, no longer are readily available in distributors' stock. This said, not only the devices and the mounting straps but the cover plates still are manufactured by P&S/Legrand. Here's a link: http://www.legrand.us/Search.aspx?q=despard

I even googled an online source for the cover plates, including not only the old standards (Figs. 8-21, 8-22, 8-23), but goof plates and plates combining Despard and standard or decora devices, such as GFCIs.

Like the exotic switches that independently control three devices from one yoke. I don't find such items worth keeping in my truck. If a customer wants one installed, I can get it, just as I can get a wall switch

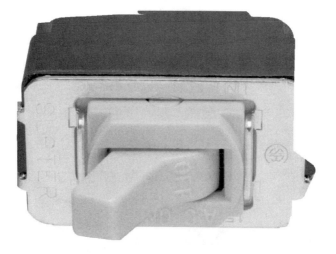

FIGURE 8-20   Despard devices included switches like this.

FIGURE 8-21
Despard covers
were unique.

FIGURE 8-22
Two-gang Despard covers only fit two Despard devices.

FIGURE 8-23
Three-gang Despard covers only fit three Despard devices.

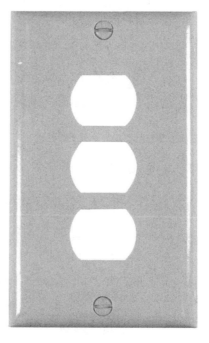

FIGURE 8-24   Not quite Decora-style devices use special covers.

with ON and OFF pushbuttons; that's enough to know. However, knowing that a replacement will not be an off-the-shelf item is enough to make it more sensible, to many customers, to go a simpler and less expensive route.

One type of device that I simply have not seen available—neither the devices nor the covers—are the precursors to Decora devices, shown in Fig. 8-24, which simply were slightly smaller rectangles. Whether the device dies or the cover is damaged, I replace both. Another is the crowfoot, also discussed in Chapter 6. The same is true, in many cases, with downstream-grounded heavy equipment, where changes have made it no longer legal.

Mercury Switches

Another old item that is mighty uncommon nowadays is the mercury switch. Early snap switches, taking after the two early ones in Fig. 8-25, did "snap," loudly. Consequently when a quiet switch was desirable, an installer used one that made and broke contact at the boundary of a small pool of nicely conductive, and noncorroding, mercury. There were no springs to fatigue, no air-exposed contacts whose limited contact surfaces would pit due to contact resistance, and eventually burn up. They last!

I emphasized their durability, because more than once, I have found one blamed for lighting malfunction when the switch actually was in good working order. Any monkeying with a mercury switch and it may malfunction, even though it is not defective. If it was reinstalled upside down, or even significantly off plumb, the pool of mercury may flow away from the contacts. Straighten it, and I have "repaired" the switch. Fortunately, I haven't needed to dispose of a mercury switch. I'd have to treat it like a fluorescent, and for the same

FIGURE **8-25**   Early snap switches looked little like modern ones.

reason. This, frankly, is the only significant reason that a mercury switch—or thermostat—is worth special mentioning.

Antique Dimmers

High-low dimmer switches offered only two levels of light. They have no serious advantage over modern full-range dimmers, and can be easily replaced.

Rheostat-type dimmers were far more rugged than modern semiconductor units. They were not destroyed by power surges or sags, such as can be caused by a circuit tripping. They could be used, with some limited degree of impunity, to control motor loads. On the down side, they also were woefully inefficient, throwing away unneeded power as heat. I haven't run into one in years.

Rebuildables

I haven't seen a rebuildable device for years, either. If I ran into a worn-out receptacle whose wearing parts could be unscrewed, I still would toss it.

Rosettes

According to consulting engineer Fred Perilstein, many ceiling fixtures were installed in Philadelphia without even shallow boxes (Fig. 8-26), as late as perhaps the early 1940s. The "rosettes" used to terminate the cables and support the fixtures do not meet with the fixtures or canopies to form complete enclosures. I will not replace such a fixture without adding a box, although I might rebuild it if possible. There is an exception: if the ceiling were combustible, I would not rebuild it without adding protection. If I were consulted, I would urge the owner to redo the setup even if the fixture were working fine.

FIGURE 8-26   Some old BX boxes had built-in fixture support.

I've never noticed one of these while inspecting, but if it had not been disturbed, I expect that I would accept it as grandfathered.

Suboptimal Designs

If it's doing its job, and is reasonably safe, I don't replace what's existing, except at the customer's request. When a protective device has been supplanted by a better design, I do tend to alert a customer to that fact. One example is GFCIs that were manufactured to the pre-2006 standard. Another is AFCIs that are not rated to protect against both parallel and series arcing. On the other hand, I will test devices with TEST buttons, and strongly urge their replacement if they fail the self-test. Smoke alarms are one exception, in that I don't always choose to make them blare out.

Grounding and Bonding

Iffy Wiring Methods

While electronic devices offer valuable protection, grounding and bonding remain very important. Despite this, even if old, ungrounded NM is in good shape, I am not inclined to augment the cable with an external ground. One reason is that it usually is almost as easy to pull in fresh cable, given that either needs to be protected. I would consider it unacceptable to stuff a ground wire behind baseboard, or to tack it like phone wire. Another reason is that water pipe grounds are no longer legal. In fact, I would recommend that customers with water pipe-grounded cable replace it.

At one time, NM, even 14 AWG NM, was legally manufactured with a reduced-size grounding conductor. So long as it's otherwise okay, I don't worry.

Old BX is another matter. Until a late-1950s standards change, BX lacked a bonding strip (although the concept was patented in 1930!). Consequently, the old stuff has high inductive impedance. Fortunately, in many, many cases unassisted cable armor has proved to successfully operate overcurrent devices by carrying faults. However, I have heard trustworthy reports of wood being charred around holes through which such cable passes, at least when the cable was deteriorated, or not correctly protected at its ampacity. This indicates that the cable got dangerously hot in response to faults that did not cause quick operation of the overcurrent device, Figure 8-27 shows such a case where the conductor insulation died while the wood charred. I understand that the assumed connection between a phenomenon

Figure 8-27 Pyrolization can char wood and destroy insulation.

Figure 8-28   Lead sheaths can tear or corrugate with flexing.

related to pyrolization, known as "pyphoric carbon," and fire is still under investigation, although not proved. However, if shavings or flammable rubbish had been at hand—one example is a vermin nest— that could have served as tinder. The problem with old BX having high impedance may be less true with the briefly manufactured double-wrap BX, which contained both a concave and an overlapping convex spiral of armor. This has not, to my knowledge, been tested. Similarly, lead-sheathed BX contained a great grounding conductor right under the armor. Lead-sheathed unarmored cable (Fig. 8-28) also ensured great grounding continuity, except in this case—because the enclosure (Fig. 8-29), contains not only the lead-sheathed cable and BX, but also newer NM with the grounding conductor loosely wrapped around the cover-mounting screw.

Even when the impedance is low due to one of these unusual designs, retaining old cable has three serious drawbacks:

- The conductors' thermal rating is low.
- The insulation is considerably bulkier than modern insulation.
- I can't expect the insulation to last as long as new THWN.

John E. Sleights researched old BX, publishing an article in the Fall 2010 *Fire Journal* about the fault-clearing ability of circuits run through BX manufactured without the bonding strip. His finding: most armor of even old 12-2 BX will exceed the temperature rating of rubber insulation when carrying a 15-amp fault or other ground

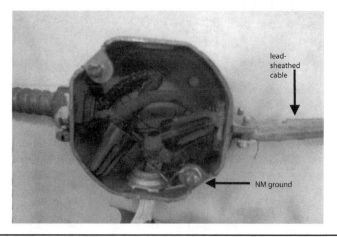

Figure 8-29 A bad NM bond interrupted excellent grounding.

return—and therefore, never blowing the fuse. John takes this all very seriously. He's seen what happens when cables aren't stopped from overheating. He spends his days as a professional forensic investigator and he spent more than 20 years in the fire service as a volunteer firefighter, an EMT, and a fire service instructor.

The Isolation Option

When I install a switch, if I do not ground it with a bonding wire, I like to use a nonmetallic cover plate. If the box and cable lack a grounding means, yoke contact is not protecting users, so I also like to use nylon 6-by-32 screws. Inspecting this, I certainly would not require nonmetallic screws.

I have seen many cases where someone has tried to terminate a wire to plastic, when conscientiously using an old-style three-prong cheater. "Conscientious, but ignorant," is not uncommon. For the same reason as I will use a plastic cover plate, I like to replace an ungrounded light fixture with a porcelain or other nonmetallic lampholder even where users are considered protected by isolation. So long as the porcelain does not have a metal pull chain, it's isolated. A brass lampholder, especially pendant, poses greater shock hazard.

I have simply replaced worn-out ungrounded receptacles with new ungrounded receptacles, but, in recent years, I've seen no jobs where it didn't make sense to add GFCI protection instead. I do have to add the "No ground present" sticker, and also add stickers to downstream outlets fed via the LOAD terminals. Unfortunately, I've just about never found such stickers on visiting a job, whether I or someone else had done the work. Not very permanent.

Example   Here is an illuminating story about the issue of affixing warning labels. After a Field Investigation by an independent listing laboratory, among other things, the laboratory's engineers decided that, before they would affix their Field Listing label to the equipment, permanent markings had to be attached, warning of the presence of two power sources. These, they specified, had to be plastic plaques engraved by professionals and they were not cheap. The permanence of such plaques made the NRTL's insistence on them reasonable.

Often, I find that a previous installer went to some other grounded circuit to pick up a ground, although this option is not explicitly permitted by the NEC. The problem is this: When permanently deenergizing part of a circuit, when I have not actually torn out the cable or conduit, I have to leave its ends accessible; this reduces the chance that I have disconnected its grounding. When I do tear out wiring, I follow certain practices:

- I make sure I know where everything leads, and get both ends of all cables.

- If any active enclosures remain, I seal any openings left in them by the cables I have removed, at least by tightening squeeze connectors or clamps.

- When I remove connectors, nipples, or conduit, I use KO seals.

- When I remove a cable from a nonmetallic box that has no clamp, I replace the box.

- Finally, when I eliminate wiring, whether because it is decrepit or because a wall is being taken out, I keep on the lookout for extra wires such as may have been installed to ground another circuit—or even to ground a telephone system. Until I've traced and dealt with these, I haven't completed demolition.

Efficiency Improvements

One area that belongs under the general category of "adapting" older electrical systems is improving the energy efficiency of a unit, building, or campus. Because there are few aspects that specifically concern old systems—luminaire heating and inrush current are two—I have relegated most of this to Chapter 7, Commercial, Institutional, and Industrial Wiring. Note especially the section title "The SWD requirement," which affects circuit breakers used as the sole controls for lighting.

CHAPTER 9

Tearing Out Old Wiring and Rewiring Older Buildings

I n this chapter I talk about what can happen when I recommend rewiring, and consider the options for doing so as elegantly and, especially, as painlessly as possible. Here "elegantly" means efficiently and well functioning; aesthetically satisfying, too, but that judgment lies with the customer; it's not my first consideration.

Property owners have various reasons to authorize—or to avoid or defer—major rewiring. I've run quite a few columns in *Electrical Contractor* reviewing what current research says about helping customers make good choices. That information, however, is not applicable only to old work, so I won't cover that ground here.

The need for total rewiring often is less than clear. One of its secondary advantages is that by disabling all the old wiring, I am sure to eliminate leftover items that never got disconnected, such as the luminaire in Fig. 9-1. Another advantage is that total replacement is more certain to catch all the deteriorated equipment, such as the wireway in Fig. 9-2, which had moisture entering through the rusty hole drilled on top for the GEC, right under the streaks where moisture dripped down the stone above it; Fig. 9-3 shows a different section under wet wood. Figure 9-4 shows other at-risk equipment in the same part of the installation. Figures 9-5 and 9-6 show problematic aspects of the same piece of work that would not otherwise have drawn attention. Redoing the whole run means that its double fire hazard is removed—the sneakier hazard being the arrowed attempt at bonding, highlighted in Fig. 9-6. It probably was a cable ground that could overheat in case of a fault, rather than causing the OC device to trip.

Example: A Judgment Call   Here is an example where the problem was, primarily, unsatisfactory specifications and design. A large, *new* house

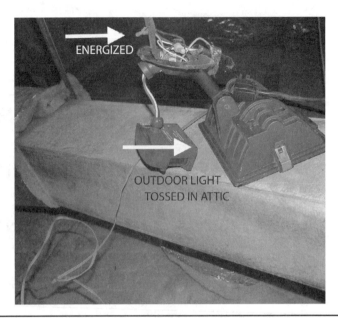

FIGURE 9-1 I watch for energized throwaways.

FIGURE 9-2 Water dripped over a stone above a GEC hole.

FIGURE 9-3   Water also dripped over wood onto this wireway.

was wired in 14 AWG NM cable, except for small appliance circuits and large appliances. The 15-amp branch circuits combined receptacles and permanent lighting, and ran as long as 100-odd ft. Voltage drop? Considerable.

Fortunately, the contract specified *satisfactory* operation of all home systems. Blinking and dimming lights were quite unsatisfactory to the purchasers. I, and subsequently an NRTL's technician, found

FIGURE 9-4   Electric rooms should be dry.

FIGURE 9-5 Cable and KO are out of the way and easy to overlook.

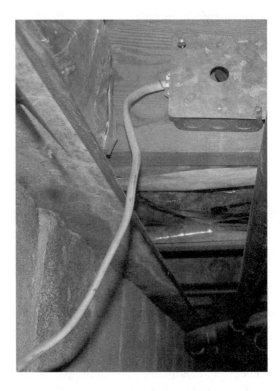

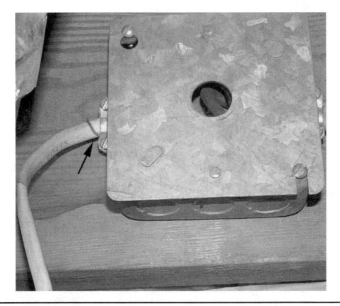

FIGURE 9-6 Up close, I also see bad bonding.

additional problems, including NEC violations that required remediation. We agreed that there was a chance voltage drop would be reduced significantly if I replaced the receptacles (all daisy-chained) and pigtailed higher grade devices.

The builder's insurance underwriter agreed to this, which cost a few thousand dollars. It helped. However, it did not help enough to satisfy the homeowner. Then I noted that the heat pump was underwired; I corrected this. Still not enough.

It took another year or so until we finally determined that the whole house needed rewiring if the lights were to stop blinking and dimming. I had to upgrade from 14 AWG to 12 and occasionally 10, and split the circuiting. The walls and ceilings had to be gutted. (In doing this, we also unearthed overlywide joist spacing.)

Safety note

I had specified the need to gut the structure, in order to rewire it efficiently. Then we saw that, not surprisingly, every outside wall was packed with fiberglass batts. The customer asked whether I wanted this removed as well. Embarrassed for not anticipating it, I said they needn't have the demolition team back. Foolish choice on my part: unhealthy and unnecessary.

When Is Rewiring Critical?

The time to recommend rewiring is when I discover how bad the wiring is.

Example   I was hired to repair a flickering, slow-starting fluorescent fixture in a kitchen ceiling. Its ballast needed replacing, and, because the fixture's original installation had been so far from legal, I spent an hour correcting its wiring. In the course of redoing the connections, I cut back on cables containing rotten insulation.

Did the house need total rewiring? I could not know. I did warn the customer that other ceiling fixtures could be in equally bad shape. Should I invest the time to check them, I asked. No, thank you.

A few weeks later, though, they had to call me back because wires were shorting at a switch box. I discovered that even there, at an enclosure which was not located near a heat source, insulation was so bad that an incoming conductor had shorted to the cable connector. With this, I knew beyond reasonable doubt that the system was shot. I did tape these wires to get this piece of the wiring marginally usable, but I also told the customer that she needed her house rewired. I also noted this on the invoice/receipt. While I knew she could not afford rewiring, I had an obligation to warn her.

Example   Another customer called me in to cure a receptacle that had gone dead. I traced the circuit up to a light fixture, where I found rotten insulation—but nothing broken. All conductors were continuous. I mentioned the condition of the insulation to the customer and kept hunting. The second light fixture that I examined was just as bad. One of the wirenuts had burned off from a splice, but not shorted to ground. The third lighting outlet that I examined also revealed conductors with rotten insulation. Here I finally tracked down the bad splice that had de-energized the receptacle. The connector was completely gone, destroyed. I told the customer that the wiring needed to be torn out. I offered to put the four outlets back together, but warned that the best that I could do in restoring them short of replacement would not make them safe. It would be far safer, I said, if I capped the wires and left them hanging until the customer was ready for rewiring. Leaving them dangling—even if replacement was delayed!—seemed far safer than torturing the brittle wires further by stuffing them back into their overcrowded boxes. Killing power to them would have been far better, but the option was unacceptable to the homeowner.

Options for Rewiring

We briefly discussed different approaches to rewiring. The customer had been considering making some architectural changes in his old house anyway. Did I have any idea what it would cost to fish in all new wiring? I gave the customer a ballpark number that curled his eyebrows. "Out of the question."

I then suggested an option that made more sense to the customer. Given the advanced age of the house, other systems probably were decrepit enough to make it worth his while to open the walls and ceilings for general upgrading. This would make my work faster, hence cheaper, than fishing. Those are not the only benefits.

Mimicking New Construction: The Gut or Semigut Job

Once the walls and ceilings have been opened, the job looks more straightforward, a whole lot like new work. In the best of cases, the demolition is handled professionally, meaning that the hazards sometimes threatening electrician, inhabitants, structure, and contents associated with opening old walls and ceilings are properly addressed. I include both the overt and the subtle dangers mentioned in Chapter 2, Hazards and Benefits.

Often, gutting enables me to create a logical, straightforward arrangement:

- One circuit per room
- More than one, in fact, if there are special needs

- Dedicated circuits, in particular, for special loads
- Properly supported paddle fans
- Intercommunicating smoke detectors hard-wired throughout
- Modern telecom wiring in the walls, if desired

In early 1996, the Electric Power Research Institute reported a study they performed, which involved sealing leaky ducts and plenums in the Pacific Northwest. In the houses they studied, most of the ducts were in uninsulated spaces. Still, they achieved a 43.5% decrease in electricity usage by the HVAC system. This should impress even customers whose ducts run within interior partitions.

Unfortunately, this customer had already paid for major work, drywalling over the old, cracked ceilings—pancake boxes screwed to joists were recessed about $3/4$ in—and decided he did not want to "undo" it.

Cooperation and Coordination with Other Trades

Illuminated optical fiber scopes are getting better and better, and the price of thermography equipment certainly has gone down incredibly in price. Still, for cases such as when a cavity is full of insulation, I've found no substitute as good as getting inside. Opening walls and ceilings is very useful—efficient compared to fishing, and often informative. However, it can be rejected as too costly. I've also found it worthwhile to inquire whether another trade will need to cut a chase or opening, with adequate, suitable room for wiring. By "suitable" I mean not snug with a hot water pipe. Given what was visible of this last house, for example, the plumbing undoubtedly was rusty iron pipe, overdue for replacement.

Lucky Breaks: When the Existing System Makes My Job Easy

I don't ignore the obvious: if there is a raceway system in good condition, my customer may be in luck. Even if its existing enclosures are undersized by present-day standards, there are solutions. I have attached extension boxes; even surface metal raceway extension boxes have been acceptable to some customers. Sometimes I can yank out old boxes and muscle in new, larger ones without distorting them, even when the old boxes had rigid metal conduit attached to opposite sides.

Such luck normally gets me only so far. A major drawback of reusing the old raceway system is that, even if I extend it, I am stuck with the old wiring layout, at least as my backbone. Furthermore, most such systems combine pipe and cable. Therefore, this option, while very attractive, almost always has to be augmented. Figure 9-7

FIGURE 9-7   Adds can make old, crowded j-boxes worse.

shows good conduit in a run I'd rather not replace—plus tired cables, one of which has been messed with by someone ignorant. Figure 9-8 shows a similarly crowded box, whose crowding was also made worse by the relatively recent addition of another cable. There, at least, the box is not part of a conduit run. I rarely want to trust existing cable when I am promising a rewire. (I never take on "just finishing up" what some stranger has begun; all too often, I've found evidence of ignorance or indifference, as shown Fig. 9-9.)

> Whether or not I am doing major rewiring, sometimes I need to deal with these same issues, as when boxes are way overcrowded, conductor insulation is shot or even half-shot, and I need to both replace wiring and add volume. Figure 9-10 shows an example.

Toughing It Out—More Difficult Approaches

The other approaches to rewiring fall into two main categories. Without removing large amounts of finish or reusing raceways, updating concealed wiring means either fishing or moving the wiring into the open. (Occasionally wireless options can replace control, voice, and data lines.)

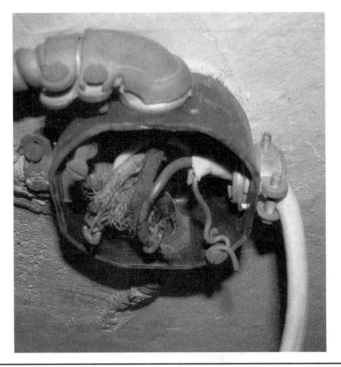

FIGURE 9-8 Here again added NM made things worse.

Fishing does not always require holes that will need patching; sometimes I can finesse access. Madisons plus box ears are no substitute for securing to a structural member, in my opinion, not even when I supplement the ears with a C-clip (Fig. 9-11). The clip is a good supplement to Madisons, I'll grant. (The concept of supplementing

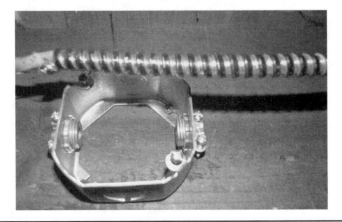

FIGURE 9-9 I can't assume what another started is suitable.

FIGURE 9-10 Conduit can make a j-box hard to replace.

FIGURE 9-11
Modern yoke/ear
extenders are
handy.

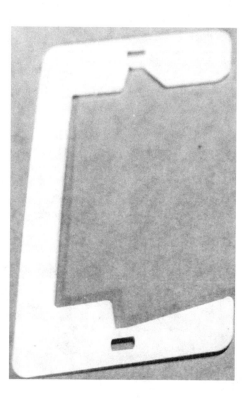

Figure 9-12   You see individual, removable yoke extenders.

yoke ears that look inadequate to bear against the wall is not new—
Fig. 9-12 shows this.) My same distrust applies to self-holding boxes,
and I've used a wide assortment of them, not just the styles shown
in Fig. 9-13. I'm a little wary even about locknut-less connectors. One
old version (Fig. 9-14) an NM connector, could not be yanked out of
the box. With the new ones, as with the old BX connector in Figs. 9-15
and 9-16, I have to trust the product standard over my experience.
This is not to say I won't use self-holding boxes, but they're far from
favorite. I am extremely unlikely to cut out an existing switch box with
a Sawzall™, substituting a self-holding box, to gain access to the wall
cavity. Therefore, even when I don't need a hole for the new enclosure,
I normally need one outside the existing box to gain access to its KOs
with a connector.

Figure 9-13   These are two of many self-holding designs.

FIGURE 9-14   A strange, setscrew NM connector.

There have been some very rare exceptions, where I've removed a KO and fished cable from above with connector attached. I've even temporarily removed a cable strap, pried out the unused cable opening, and fished in a cable from above. These fishing stratagems can take a lot of time.

I am greatly impressed with the new boxes that allow me to cut away a switch box secured to a stud, fish NM cable to the resulting hole, and install a new box that is Listed for screwing to structure from

FIGURE 9-15   The outside of an odd old BX connector.

FIGURE 9-16   It didn't use locknuts.

within (Figs. 9-17, 9-18, and 9-19 are some examples). I even stock retrofittable joist-mounted boxes Listed for fan support, both non-metallic and metal. Way too often, unfortunately, I've seen a standard new-work NM switch box retrofitted into an existing wall, dry-wallscrewed to a stud, the screw heads illegally sticking into the box (Fig. 2-6).

FIGURE 9-17   I prefer to retrofit boxes that screw to studs.

FIGURE 9-18   I don't care that much which design I use.

FIGURE 9-19   I'm delighted they come in multigang styles.

None of this gets me beyond the cavity created by a pair of adjacent studs, unless there's a balloon wall. For horizonal travel, I have pulled away baseboard (or, in very rare cases, wooden crown molding), when it does not appear brittle, to open the wall and drill studs and joists behind it. If my drilling angle makes it necessary, I add nail plates. This is a step that I find installers often overlook. Diving down to an unfinished basement or up to an unfinished attic, I've run exposed cable inoffensively, through a less-habitable space. Sometimes exposing it inside closets or pantries is an equally acceptable cost-saver. When I do this, I have to make sure it is not likely to be subjected to mechanical damage.

Note
Experience has taught me how common and easy it is to for occupants to abuse exposed cable.

Even when I drill joists in such spaces, I'm doing less aesthetic damage that requires patching. I am careful not to weaken joists by chewing them up too much. Section 2517 of the Uniform Building *Code* restricts hole diameters to $1/3$ the depth of a joist.

When I do cut into a wall or ceiling, in older buildings, often I am not dealing with drywall. I have patched plaster or gypsum block walls using plaster of paris or patching plaster, with wire lath to support larger patches. Many times, the customer is happier for me to leave this to a professional, or at least someone not charging electrician's rates. The biggest problem with plaster ceilings is that they can be relatively brittle. If too much ceiling comes down, I have to leave repair to the plasterers.

Some old-timers swear by the option of cutting through the tongues of tongue-and-groove wooden flooring and temporarily removing a plank to run cable through floor joists. I have not tried this.

Fishing varies in difficulty. Fishing down can be very easy if a builder used balloon construction, skimped on firestops, and didn't insulate. However, this is precisely where the ease of long fished runs sometimes results in problems such as shown in Fig 9-20, where the cable—fortunately BXL, whose inner sheath maintains grounding continuity—pulled out of the box connector because it was not secured nearby (even supporting is not the same as securing) and had a considerable amount of weight pulling on it. Normally, though, I run into plates at the top (and bottom) of each wall, or firestops.

These are far from the only obstructions. Even though not as solid as blankets, blown-in insulation can make fishing more like ramming. Then there are the wads of newspaper that previous workers have

Figure 9-20 When BXL's armor pulls out, the lead may not.

stuffed into holes to make them easier to plaster over, and occasional buried trash. Last, but far from least in importance, are HVAC ducts, plumbing pipes, and other wires. I've been extremely lucky not to damage these. What I have done, unfortunately, is drill up into a floor or a roof. Fortunately, the holes were small, the roof was asphalt shingle, and it was easily patched.

Everybody has their own system for fishing. There are systems without number for locating, ranging from stud/joist finders—useless when the ceiling is too thick—to banging, to sophisticated and expensive electronics. New patented tools come along every so often. I've learned to measure and remeasure, to work from finished space to unfinished initially. After my preliminary effort, I sometimes drill a $\frac{1}{8}$ in hole down between the trim molding and the edge of the baseboard into an unfinished basement to confirm that the location I measured is where I think it is. I can then move a few inches out and drill up through the plate into the wall under the receptacle, sconce, or switch.

Carpeting will hide a small "locator" hole, but carpet can grab a drill bit. Steve Jones of Jones Electric told me he used a piece of wire coat hanger as a bit, drilling with light pressure. I've used a twist drill, run backward so as to avoid winding fibers around its grooves.

I'm aware that I am NOT using the bit in the manner for which it was designed.

Example: Going Fishing

To bring this down to earth, here's a typical rewiring job in a structure that calls for fishing rather than demolition. It could be historic, or toxic, or in continual use, or just owned by someone who wants the work done this way despite mess and inefficiency. The building is an occupied single-family residence, with two fully habitable stories, an unfinished attic, and a partly finished basement. The service equipment is in the basement, which is substantially below grade level.

Further Choices

Leaving the Existing Wiring in Place

With instructions to keep mess to a minimum, I may blank off the old outlets rather than removing them. This represents a real trade off in terms of aesthetics, as well as in terms of potential confusion. On the other hand, it has the advantage that I need not touch the old wiring until my new system is in place and tested. The main exception is where a new outlet simply must be located in the identical place the old one occupied—for instance, the center of the ceiling in the case of a lighting outlet. An old lighting outlet can be extremely crowded, making it especially valuable to replace the box with a much larger one.

Only in rare cases has it made sense to try to reuse the old boxes with new cables. For one thing, doing so necessarily requires pulling out the old cables, which usually are stapled. Left in place, they would occupy needed box volume and also present too great a potential for mix-ups. Furthermore, old boxes may not have tapped grounding holes, they often are unnecessarily, even illegally, recessed, and sometimes the boxes are damaged.

New boxes can provide just the type of internal clamps I want, plus all the volume I might need. Abandoned, the old boxes can sit there peacefully, with blank covers, until the building falls down. If someone does decide to bury them, at least abandoned boxes no longer are connected to the electrical system. If the old wiring at no point enters an enclosure containing energized wires, it even is arguably legal to plaster over them. However, if any lines still enter them, all it would take is one overlooked connection between the discontinued and the new wiring, and I would have a potential hazard—not to mention possibly a real puzzler when troubleshooting.

Of course I'll reuse a new, appropriate disconnect, such as Fig. 9-21. But even though the wires are not old, the ignorance shown by double-tapping the line side lugs without a pigtail, even if double-tapping

FIGURE 9-21 A pigtail's easy; some violations must mean ignorance.

them were suitable, means I wouldn't consider retaining the earlier installer's wiring.

Removing old cables is the best choice, from a technical standpoint. Not only does this reduce the chance for confusion, but it removes a potential fuel from fires—a fuel that could give off quite-toxic smoke. On the other hand, it does take time, which adds cost.

Unobtrusive Cables

Sometimes, I permit the old system to dictate the design of the new, at least in part. To remove all the old wiring, I have to open walls and ceilings to so great an extent that I try to use those same routes for most of my new wiring, rather than make yet more openings. Sometimes, I can reuse the existing holes through building members. Old cables tend to be fatter than modern ones. I do try to minimize voltage drop, labor, and mess.

> **Resource**
> The July 2009 issue of *IAEI News* has an article by Travis Lindsay on selling upsized wiring runs to residential customers by calculating Return-On-Investment from reduced voltage drop. For a parallel least-cost analysis oriented toward commercial customers, search for the Alcan-sponsored 2007 conference presentation, "Energy Efficiency in Design of Electrical Building Cables," by Paolo Kistler, Gerald Rebitzer, and Ravindra Ganatra.

More Detailed Examples: The Up-or-Down Dodge   The next section will take the ideas I alluded to about maximizing fishing convenience, and make them more concrete.

Example   I rewired a top-floor bedroom, a 10 × 15-ft. rectangle with one entrance. It started with one receptacle on each wall, and a pull chain light in the middle of the ceiling. When I finished, it had eight receptacles and two wall switches. One switch controlled a light fixture in the center of the ceiling; the other controlled half of a split-wired receptacle. I also installed a dual-mode smoke alarm, intercommunicating with the alarms elsewhere in the house. At that time, dual-mode, dual-source smoke alarms were not available. (As this edition is being finalized, they have just reentered the market, after an earlier version was recalled.)

This bedroom was on the top floor below an accessible, unfinished attic, with no firestops in the walls, so I could fish without making additional wall openings beyond those required for the new boxes. While I could loop from one new outlet to the next through the attic, protecting the cable up there as needed, I prefer to run all of them from one attic junction box, reducing voltage drop. I could have run cable up and down from receptacle to receptacle to switch to light. This has the trivial advantage that all splices are right there, accessible without leaving the room. Either way, I would bring it all home to one new circuit on one home run, if I were eliminating the old receptacles. This wasn't authorized here, so I ran from a replacement ceiling outlet box. If I preferred lighting to be on a separate circuit from receptacles, this too would have been easy to arrange. The smoke alarm cable traveled up and down through the attic from one top-floor alarm to the next.

> **Note**
> I do try to keep equipment out of thermal insulation, whether the insulation is present or likely to be added in the future. Sometimes, I create a dam around a lighting outlet.

When a bedroom is on the first floor, over the unfinished portion of the basement, I often accomplish something similar by looping through an unfinished basement, down and back up from receptacle to receptacle to switch. I do often have to chop holes in order to fish cables to ceiling outlets.

> **Warning: Special hazards in crawlspaces**
> Then there is wiring in a crawlspace. Besides their requiring contortion, and permitting limited vision, I've run into—more like crawled up to and sometimes on to—bugs, rodents, pieces of brick, broken glass, and jagged metal in crawlspaces. I like to bring a tarp.
> I've also gotten tingles, especially in under-the-building crawlspaces. So far, nothing more serious. A kludged connection or a splice

whose connector has fallen off can be deadly. I am likely to be well grounded and too often I'm wedged in, restricting flinching that might pull me away if the contact were poor enough. At least I don't work in shorts or artificial fabrics or wear metal. But I could do better.

Because crawlspaces rarely if ever see an inspector, they are more likely to sustain such electrical hazards. If new wiring dives into one, if I'm the inspector I want to look at it. (That was a peculiar use of "want to.")

Examples: Doing It the Hard Way   When I rewire finished rooms that are located both under and over finished spaces, there's a trade-off between the rational and the easy. I cannot loop back and forth from an unfinished space unless I am willing to create a not-very-rational wiring layout. Therefore, unless surface wiring is an acceptable option, I usually face the tedious and messy work of drilling through every stud between one receptacle and the next.

Here is an illustration of how I've compromised logic for up-front savings. A first-floor room is right under a second-floor room, with an unfinished attic above that. The house is built on a slab. I've dropped cable from the attic down to a receptacle in this second-floor room, drilled through the floor plate under this second-floor receptacle, or, if the receptacle is in the outer wall in balloon construction, just dropped down. With no obstructions, I now drop down to a first floor outlet location without any further drilling—certainly without drilling horizontally through the half-dozen studs. I hate this approach. Many, many old houses have circuiting much like this, where the odd receptacle was added by picking up power from an outlet upstairs or downstairs.

This approach is hell on future troubleshooting, not to mention destroying any chance of creating an intelligible panelboard directory. On the other hand, if the square footage and loads are such that it makes sense to have most of the habitable rooms on one circuit, it is generally not harmful or illegal. It even can make sense, say for a special circuit serving several very small loads. I won't ignore *Code* rules, but I will loop cable along roundabout routes to get rid of extension cords.

New wiring does have to meet the circuiting requirements in the currently enforced version of the *Code*. No inspector has demanded I change a legacy layout when merely replacing a length of bad cable. With substantial rewiring, I have to bring the system up to date. I suppose there could be grounds for a special pleading, but I haven't faced this situation. There's more on this in Chapter 11, "Grandfathering, Dating, and Historic Buildings."

Chopping or Disassembling

When I see no way to avoid lots of drilling, I have to damage or at least pull apart some portion of the structure. I have a choice of procedures.

Paneling Can Offer a Less Obtrusive Option

There are a few additional ways to reduce the tedious fishing required when a customer does not want walls and ceilings ravaged. With wood paneling, sometimes I can remove a whole sheet of paneling and, after doing what is equivalent to new work, restore it. The best old paneling, though, is constructed from tongue-and-groove strips, like wide floorboards. No short-cut there.

My fallback approach when working around new-enough walls is cutting away a strip of drywall and replacing it after drilling and wiring behind. With luck, I can hinge it out from the surface, rather than removing it completely. Sometimes, this will let me reattach it without having spanned two or more studs.

In and Out

The next system is at the same time very ugly and, in certain setups, quite discreet. Whether I am trying to get power from a basement to a higher story, or from receptacle to receptacle along an outside wall, sometimes a fast and easy way to go is to hide it outside the building. (This is a bit like using a crawlspace.) The appearance of miscellaneous wiring on the front wall of a house could be unacceptably offensive. Sides that abut blind alleys are more promising. In these cases, no one may care if I neatly secure RNMC, EMT, or UF on the outside of the wall. With a factory, the wiring actually may be safer installed outdoors than it would be if run from ceiling to floor in the same factory as NM-B or UF-B, with a minimal 6 in of mechanical protection as it emerges from floors.

I've run cable alongside downspouts, carefully following the exposed wood of timber-and-stucco houses, and above window frames.

Securing   Chapter 10, "Inspection Issues," talks more about installing supports outside that are adequate technically, as well as aesthetically. One final matter. While working outside to secure a raceway or cable that penetrates walls, I will carefully caulk around its entrance and exit to bar moisture and vermin. Putting aside structural damage, I have found any number of panelboards to need replacement that might have survived further decades if not for inadequate caulking. In Fig. 9-22, aside from exiting a sharp edge from its EMT sleeve, the cable entry is uncaulked: unacceptable, despite the overhang.

FIGURE 9-22
Sleeves require
smooth ends or
bushings.

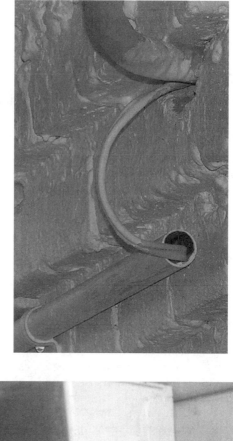

FIGURE 9-23 Here's the outside of the misapplication.

Figure **9-24** NM can attach to a Wiremold® box—with a connector.

Exposed Interior Work

Surface Cables and Raceways

The alternative to leaving things alone, burying, or hiding my work is exposing it. I always show a sample of the system we plan to use to any customer who may be unfamiliar with it before getting started. The least expensive but most unsightly such option is simply running cable along the surface. This is almost never authorized in a bedroom or family room, but I've installed it in many a workshop, in some commercial and even home offices, and in any number of "nonpublic" areas. I do want to make quite sure that it's not going to be subject to mechanical damage. Often I'll insist on BX or UF. EMT is not a whole lot more expensive to install and can hold up yet better. Sometimes a customer specifies EMT; fine. Wiremold® can be convenient, especially for transitions from concealed wiring. While some customers like its tidiness, if it's at risk of being bumped—and it rarely isn't—I prefer to use two-hole straps, despite the fact that these are a bit less tidy-looking.

Inspection note
When looking at other people's rewiring, I keep very alert to misuses of Wiremold®, such as Fig. 9-23 and, less obvious until it was taken apart, Fig. 9-24. I am also aware, though, that a century ago the

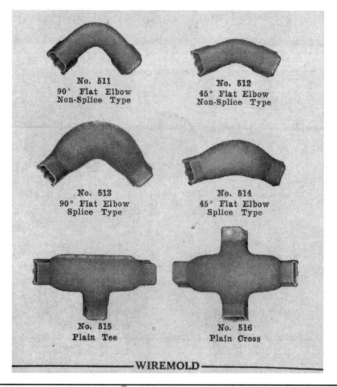

FIGURE 9-25 1917 Wiremold® fittings were just a little different.

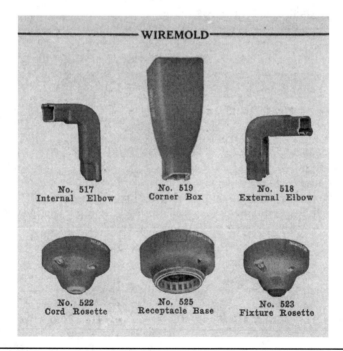

FIGURE 9-26 In 1917, they did make canopies to serve fixtures.

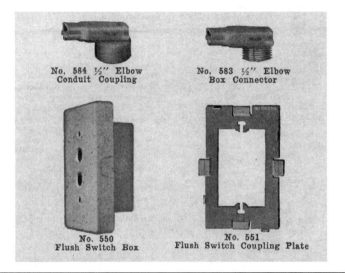

No. 584 ½″ Elbow
Conduit Coupling

No. 583 ½″ Elbow
Box Connector

No. 550
Flush Switch Box

No. 551
Flush Switch Coupling Plate

FIGURE 9-27   Their switch boxes were made for double pushbuttons.

raceway's fittings not only looked slightly different, but were designed for additional purposes. Their No. 513 and 514 splicing ells, in Fig. 9-25, are an example of this possible confusion. Contrariwise, the appropriate uses of the fixture canopies/rosettes of Fig. 9-26 and the conduit connector and the pushbutton cover of Fig. 9-27, similarly antique, are unmistakeable.

CHAPTER 10

Inspection Issues

Inspection Counts!

It's time to focus on my work as a third-party inspector; as a consultant, sometimes supporting home inspections or legal proceedings; and as a contractor concerned about compliance issues, including permits for and inspection of work on old systems. I'll switch back and forth between these perspectives, always focusing on inspection.

This is one of the most important chapters. Inspection seeks *Code* compliance, but ultimately is about making things safer. Countless inspection-related issues arise; IAEI (www.iaei.org) is a good resource for exploring them. I'll discuss my experience with some I've found particularly relevant to old wiring. As always, I am not spelling out how-tos, nor laying down the law.

To start, as a contractor I mostly find getting work inspected to be very much worthwhile, at least when jurisdictions employ qualified, responsible inspectors. Some multi hats (combination inspectors) try hard, and one of them might notice that where the supplementary grounding bar in Fig. 10-1 is attached hides important printing, even when Sparky overlooks it. We're so used to seeing everything blocked by wires! But only we are at all likely to check for mixed neutrals and grounding conductors in subpanels.

Departments eliminating experienced (expensive) employees sacrifice quality, and privatizing sometimes has the same effect. I still hear of unethical behavior such as bribery, or third-party agencies seeking business advantage through slack inspection. I see both as corrupt practices. Inspection should mean one more person looking things over, catching problems the installer missed. Inspection also tells the customer that I have nothing to hide, especially useful when I've warned them their wiring was messed up. All this is well worth a few dollars and some inconvenience.

On a related note, some jurisdictions offer, and encourage me to use, postcard permits that allow me to self-certify work. I do use them, at a customer's preference, to save expense and trouble. Unfortunately, I don't believe the work covered by my postcards has ever

Figure 10-1   Retrofitted bars can block needed information.

been double-checked. One exception: Washington, DC does allow the use of a postcard permit for heavy-ups as well as for trivial jobs, and the service upgrades must be called in for inspection, because the utility is not going to accept my own cut-in notice. Generally, though, self-certification means the customer has traded down to less protection.

There's no question that even the best inspection can't be exhaustive, but almost always has to be a "best-guess" evaluation. It's long ago, if not far away, that we could follow the advice in Fig. 10-2—which goes so far back that the AHJ was assumed to represent the insurance company, not the locality. Inspection still takes time and expertise—plus luck, to catch potentially dangerous sloppiness like that shown in Fig. 10-27 bare copper's, or the individual conductors, not cable, exposed *behind* the FD in Fig. 10-3. I turned down full-time inspection work for one jurisdiction when I learned how little time they allotted per inspection. Certainly, I don't believe that anyone expected to perform 50 inspections a day (I've heard of this!), can do a proper job, and even glance at the outsides of panels closely enough (Fig. 10-4), although a seasoned AHJ does develop effective shortcuts.

The inspection community is not alone in our concern about proper inspection. Insurance companies have made several attempts to sue local fire departments, citing failure to properly inspect and warn building owners of fire hazards. The Delaware Fire Marshal arrested an inspector for not enforcing state requirements, using his

Electrical Inspection. The principal points regarding the safe installation of dynamos, motors, heaters and outside and inside wiring, as required by the insurance underwriters, have been briefly set forth in this little book, which has been compiled simply for reference and not as a teacher—a book designed to settle most of the doubtful questions which might arise in the mind of the engineer or contractor as to just what will be considered safe by insurance inspectors. There will probably arise questions which cannot be settled by reference to the suggestions herein contained, and, therefore, a great deal has to be left to the judgment of the constructing engineer and inspector. In every such case the Inspection Department having jurisdiction should be consulted with perfect assurance that nothing unreasonable will ever be demanded in the way of special construction.

Every piece of wiring or electrical construction work, whether open or concealed, should be inspected, and *notice*, therefore, *should always be sent by the contractor or engineer to the board having jurisdiction immediately upon completion of any work.* Negligence in this matter has frequently caused floors to be torn up when doubtful work has been suspected, and at the cost to the contractor.

FIGURE 10-2 1917 advice said, "Inspect everything."

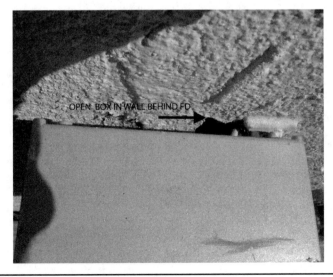

FIGURE 10-3 Caulk would have hidden illegalities behind a surface box.

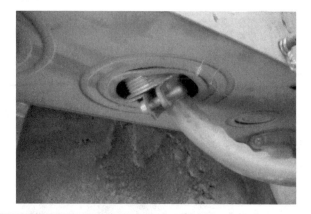

FIGURE 10-4   Some violations are right out there.

arresting authority to ensure that those who are authorized and responsible to protect the public from electrical hazards do so, according to report.

Judgment Calls

Inspectors face many questions and controversies. I'll discuss several confusing areas, starting with the phrase, "neat and workmanlike."

Neat and Workmanlike

It's easy to feel lost when dealing with old wiring. Because the "general-duty clause" of Section 110.12 is such a subjective call, it's easily abused when an inspector's floundering. Every authority agrees that 110.12 doesn't authorize "Do it the way I would" inspection. How about "If he cares this little about making the job look good, who knows what corners he cut?" I don't need 110.12 when the job patently hasn't had a walk-through before the contractor called for inspection, as was the case in Fig. 10-5, where an energized tail was lying on the kitchen floor when I arrived for the final. In less blatant cases, I may still think the contractor didn't care, but I try to inspect to *Code*. What constitutes "to *Code*" can require research, which I discuss more in Chapter 11, Grandfathering, Dating, and Historic Buildings.

Section 110.12 is there to fill gaps—places where the NEC fails to specify details. Barring local amendment, the *Code* references just an NEIS concerning workmanship, in an unenforceable Informational Note/*fpn*. Experts argue that 110.12 is there for failing an installation where something was done wrong, but where the correct approach is not specified explicitly in the *Code* because it is so obvious. I tend to agree, and I add my own perspective.

Figure 10-5   Free + energized at Final = dangerous work.

With workmanship, I try to maintain a double standard. As a representative of the AHJ, it is my duty to enforce the minimum NEC requirements (plus any local amendments), and no more. As a conscientious contractor, I need to do a job I can feel good about.

Example

In my role as a contractor, I emailed an AHJ about a forthcoming job. I had been asked to replace all the backstabbed receptacles in a single-family residence with receptacles wired in a manner generally considered more reliable. In addition to answering questions about licensing and permits, he volunteered that the way the existing receptacles presently were wired is in keeping with their UL Listing, and therefore acceptable. He may not like it, but he saw his duty to approve it—and not to leave me thinking otherwise!

> This does *not* mean that replacements could be wired identically. I've seen this mistake made by a cautiously bold Do-It-Yourselfer, who decided to replace a worn-out receptacle by religiously following the design used by the (competent) installer. Whether it's a matter of overlooking the requirement to pigtail multiwire neutrals or the restriction on backstab device connections to 14 AWG, he's not the only one who makes it valuable to inspect even such "dirt-simple" jobs.

Reasonable Accommodation

Article 90 does not say we are expected to provide an absolute guarantee of safety. Sometimes the choices that are in any way plausible mean picking between several less-than-ideal solutions. That's what fiduciary responsibility—being the nearest thing to an expert who's available—is about.

Example   On two jobs, the considerably experienced master electrician on the spot was unable to identify what loads some circuits served and had left them on and unlabeled. I had him turn them off and made sure the occupant understood the importance of labeling, why they were left off, and when to try them.

One quite-experienced inspector colleague, put in my place, would have removed the leads from the breakers, perhaps labeled them as spares, and left a record. Another, even more experienced, would have left them on, against the chance that they controlled important loads and leaving them off would have caused damage before someone realized the reason for, say, ice-melting wiring to be nonfunctional. He'd also have made sure the HO understood that they could represent some risk, and should be labeled.

Filling the Holes in NEC's Coverage

When the NEC does not provide clear guidance as to how a piece of equipment is to be secured or mounted, I have to extrapolate.

Example

To route conductors for a new circuit past a finished space in an old building, it can make sense to dive outside and bypass walls and ceilings. To guide the installation of UF cable on an outside wall, as a contractor I might use the "closely follow the building" rule of Section 334.15(A) for inside NM cable, supplemented by the 4½ ft. support rule in Section 334.30 (or Sections 338.10(B) (4)(b), 230.51 (A), and 230.54(D), all of which address SE cable). As an inspector? Nothing so specific, but not "anything goes," either. For starters, I just wouldn't allow cable to be left dangling. I also make sure the way it's secured won't let it move significantly and is suitable for the environment. The abrasion in Fig. 10-6 came from rubbing by the rusty strap in Fig. 10-7.

The inspectors who earn my greatest respect apply a rule that is at the same time looser—more generic—and tougher than good contractors. Where no detailed guidance is offered, installations must be sufficiently "workmanlike" *to avoid hazard;* installations that are so sloppy as to create risk are red-tagged—we will accept those that merely offend the eye. So long as the next electrician in will be able to work safely, messiness, as such, is an issue between the customer and the contractor.

There are, however, cases where a rule is totally ignored. I will not fuss too much about a couple of strands of a large, relatively fine-stranded grounding conductor not making it into the connector. However, in Fig. 10-8, this practice went way past dangerous. Even

FIGURE 10-6 Both UF sheath and insulation are injured.

if the HVAC installers hadn't performed illegally jumpered the lugs (with an undersized conductor of incompatible material!) they just about destroyed the grounding conductor, in addition to ignoring the cabinet's bonding lug.

Examples

If UF is likely to get torn loose, that's in my bailiwick when inspecting. If the cable or conduit is that poorly secured, I insist that sufficient straps or staples are added. Twisted? Ugly? Tsk. Not a violation, just a shake of the head.

FIGURE 10-7 Rusted straps can harm cable.

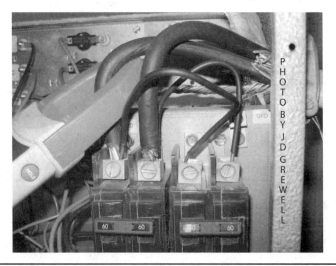

FIGURE 10-8   Lug misuse, Al and Cu and clipped wire are all common.

A customer complained her contractor had installed the panel-board upside down. The carton picture proved it! I had to explain that the picture didn't make it a *Code*/Listing issue: every handle operated horizontally.

How about the completeness and legibility of a panel directory? With the hodgepodged circuiting in old buildings, usually the line on the circuit directory intended for a circuit's legend is thoroughly inadequate. I find it impossible to fit a legible, reliable description on it of what the average fuse or CB controls—and what it doesn't control, by which I mean those items that fit the same description unless it's detailed. I'll return to this issue.

When Am I Required to Pull a Permit?

In rewiring, permits and inspection are not always required. Most jurisdictions trust professionals to perform limited repairs, sometimes even extending to replacing small rotten bits of cable. A gut for rewiring, though, generally is treated mostly like new work—which is not this book's topic.

I have never needed a permit to replace a receptacle or three. Yet "replacing" doesn't begin to describe the job. Often bonding is needed. Sometimes, I need to replace the box, because it's damaged or because it's a plastic replacement box that someone used with BX. Sometimes locating a hidden panel takes longer than the repair itself, fancy tracer or no. Contrariwise, many jurisdictions demand a permit to replace even 10 receptacles. Inspecting this, I can be fairly quick, although I do ask to see inside the boxes, and the OC device. If the installer has no idea of the circuit, having worked while the outlets remained energized, I'm inclined to leave a red tag. In some old buildings I've worked on—and not only commercial occupancies—even pulling the obvious meter wouldn't have killed power in a deadly emergency.

Now suppose the customer is redoing an old building, but upgrading room by room. Normally, where the walls and ceilings have not been opened completely (or at least to the extent that will readily allow installers to mount boxes and cables according to new wiring rules), there is only one inspection, at the end. Piecemeal inspection, before each hole's patched, would take an impossible number of visits. How does this jibe with the idea that when fresh wiring is run, it should be inspected before final energizing, for a last chance to make sure it is safe before the customer gets to utilize it? This is one reason I look inside boxes. What works best is when contractor and inspector put our heads together beforehand.

Can the Installation Be Grandfathered?

It would be nice if there were a rule book on this. *NFPA 73: Electrical Inspection Code for Existing Dwellings* uses mandatory language, but is not terribly specific. It does formalize one widely accepted doctrine: items that never could have passed inspection, such as zipcord power wiring, must be corrected—at least if they get worked on. Does an old system have to be brought up to current *Code*? Same answer. As a contractor, I point out the dangers to my customers, and offer them choice.

Example

I was asked to replace the unsatisfactory recessed lights in a rectory kitchen. I pointed out that the wiring, while not appearing imminently hazardous, obviously had been kludged; to take responsibility for even one or two new cans, I would have to correct this. Being more cash-strapped than scared, the customer opted instead to try different lamps in the existing fixtures.

To even consider grandfathering an installation, it is nice to have some assurance that what I find was legal when it started out. Chapter 11, Grandfathering, Dating, and Historic Buildings, can provide some evidence as to whether an old, odd installation simply represents the way things were done—and approved—way back when. W. Creighton Schwan's final book, *Behind the Code*, provides additional clues. (I worked on this with him, and then polished it some more after his death, with his personal library for help.) When these questions come up, I consult over 40 NECs in hunting down the answers, plus *Code* handbooks, plus more. Some consultants have far greater resources, and also have more memory of what was done when. I still face a hard call as both electrician and inspector when equipment was legal when installed, but no longer complies with *Code*.

Examples

Here are some once-legal installations that I may or may not want to grandfather. Every decision depends on the conditions found in the specific situation. Falling-off insulation is a "gimme," forcing rewiring.

Are these action okay? Here are my calls on them:

- ❑ *Action:* Replace a switch or receptacle in a now-undersized box.
  Response: Within reason.

- ❑ *Action:* Replace a light fixture at an undersized box.
  Response: Usually yes, depending on location and crowding; I'm very concerned about pinching conductors. I find that many installers overlook 110.3(B) with respect to minimum conductor temperature ratings.

- ❑ *Action:* Extend a circuit from an undersized box (and therefore add to its crowding).
  Response: Not without adding a box extension or, if the old conductors can withstand the handling, replacing the box.

- ❑ *Action:* Replace a cable leading from an undersized switch box to a lighting outlet, in order to accommodate the demands of a replacement luminaire for conductors rated 90°C.
  Response: Yes.

- ❑ *Action:* Add circuits to split up loads fed from a now-under-sized loadcenter, or a split-bus service.
  Response: Sure, within the restrictions on the panelboard's schematic; even adding circuits to a service with only 60 amps incoming and more than six branch circuits can be acceptable. What if there is no label? In this case I add circuits reluctantly, if at all. Certainly I am disinclined to install tandem circuit breakers, because it so often is true that the busbar locations at which they may be installed legitimately—if any—are limited.

- ❑ *Action:* Add branch circuits from the MAIN section of a panel.
  Response: Red tag.

- ❑ *Action:* Add major loads to an undersized service
  Response: Nope. Of course, the characterization, "major" itself is a judgment call.

- ❑ *Action:* Add a subpanel, run from the main section of a fusebox with no room for more circuits, to split up loads.
  Response: Usually.

- ❑ *Action:* Replace a multipoint three-way or four-way switch connected Carter System style (See Chapter 5)
  Response: Probably I'd insist on a safer design.

□ *Action:* Add a circuit to an accessible, energized-front, ceiling-mounted (unswitched, cartridge) fusebox that has an empty fuseholder.
Response: Maybe.

□ *Action:* Add circuits to loadcenters in other illegal locations, such as over the kitchen sink.
Response: Maybe.

I may check with the chief inspector, if available, before going ahead. This is not only to spread the responsibility but also to get a second opinion. The best inspectors work with, rather than against, reasonable, conscientious contractors.

Example

I am still impressed with the accommodation offered when a customer in a very old house wanted a paddle fan in a tiny, peaked attic bedroom. Even short blades would extend below 7 ft., or hit the ceiling. The AHJ agreed on a protective barrier. Use a decorative lattice? Fine.

Working with contractors whom I trust does not mean neglecting to inspect their work properly. I've heard, "Joe wouldn't red-tag me for this!" Well, this may in fact be true. However, on today's inspection, I carry the responsibility, not Joe. Show me in your *Code* book, your *White Book,* or even your *Handbook.* Meetings such as IAEI's are the places to work on consistency of *Code* interpretation and enforcement. Maybe Joe is the inspector who signed the TPF in Fig. 10-9, even

FIGURE 10-9 Local rules had required tucking these in the can.

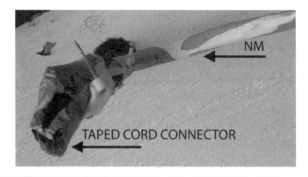

FIGURE 10-10   I've learned to check carefully under eaves.

though at the time it was signed the local rule said to tuck service pigtails and bugs inside meter cans. True, homeowner permits deserve a little extra time, to catch well intentioned, but dangerously, illegal mistakes (e.g., Fig. 10-10—and even Fig. 10-11). This means extra time beyond what's normally sufficient, rather than making an exception in allotting sufficient time to their jobs. It was more likely someone working for a contractor who stressed BX beyond its limit (Fig. 10-12). This is different from the sprung armor shown in the last chapter in Fig. 9-20, in three ways. First, the earlier image shows armor that

FIGURE 10-11   As in Chapter 4, wood is now part of the enclosure.

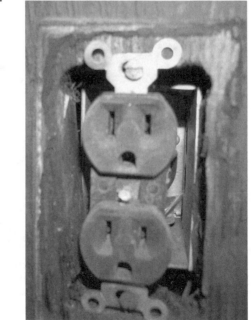

Figure 10-12 Unsupported BX by itself may spring armor.

probably was overstressed over time, conceivably as a strap loosened or came off. Second, Fig. 10-12 shows careless workmanship, in the open KO. Third, the BXL almost certainly retained good grounding continuity and conductor protection because its interior sheath.

When Is It Reasonable to Reconnect Old Equipment?

Listing compliance is another troublesome issue. In new installations I want to see manufacturers' instructions for the equipment installed—especially if the customers themselves provided the equipment. I've held up a job while we spent half an hour to chase down a manufacturer's technical people, in order to confirm that a particular use not mentioned in a Listed product's instructions was in fact consistent with their intention. The new equipment was going into a 1920s house. The salesman's assurance was not enough to convince me.

If there is no evident Listing mark, or if there is reason to suspect that the installation is not in accordance with the maker's intended use, what am I to do, as contractor or inspector, about equipment that is no longer manufactured? How shall I deal with old gear? Whether it simply is located in an area being renovated or it was removed and is being reinstalled, it is not reasonable to expect the instructions, and even markings, to be available. Even some interior labels fall off. Warning: I have sometimes found the reason was overheating, due to misapplication. I have no simple answer . . . knowing old products.

As to Listing marks, UL's the most common one found on electrical products—although it was not the first NRTL. Here's the web site where I find not only the latest versions of Listing marks but also information as to when testing agencies lost their NRTL status: http://www.osha.gov/dts/otpca/nrtl/nrtlmrk.html. (Two agencies went off the list in 2008; one for cause.) It leads to links showing the scope of products agencies are authorized to List.

Knowledge of the NEC and of accepted wiring practice is my basic area of competence, not ANSI standards. I've received a lot of assistance from UL representatives in getting at least a bird's-eye view of relevant standards. When there is serious question, absent appropriate documentation, extreme cases—including particularly suspect equipment—can demand field labeling or certification, preferably by a NRTL. (Note that a customer may judge this to be prohibitively expensive.)

After-the-Fact Inspection

I don't know of inspection departments that perform courtesy inspections. That's where I come in as a consultant. I'd love to see reinspection required, perhaps on any new property sale, lease or rental. How about high risk, public access sites—gas stations, like the one where I snapped Fig. 10-13? Sometimes the AHJ insists that a permit be pulled to cover work that was performed by a fly-by-night. Or an owner gets nervous and wants someone to come in and not only to look over their wiring but take responsibility for it. I'll take this on, though even inspecting a job on which a permit was pulled only after the installers were caught working illegally can be touchy, unless I am going to have them pull everything open, including paneling or drywall. Here's a good case for the compromise of meggering. Using megohmmeters goes back a long way (Fig. 10-14). Incidentally, I've found a fairly early megger not only to be a tough old beast but useful (Fig. 10-15 shows my backup megger).

Figure 10-13 What good are seal-offs with a box cover missing?

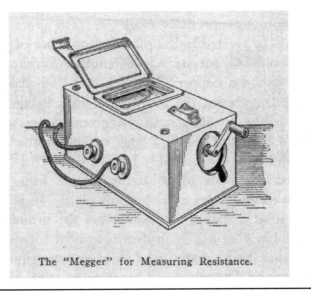

The "Megger" for Measuring Resistance.

FIGURE 10-14 Basic Megger design lasted decades after the 'teens.

Another tough job to do well comes when I'm asked for a big-picture assessment. In one sense, home inspectors have more freedom when preparing electrical reports. First, they have ASHI standards to guide them: shorter and simpler than the NEC. They must open panelboards and look inside. Elsewhere in the system, they fulfill their duty if they eyeball what's exposed and plug in a tester. Second, I can't

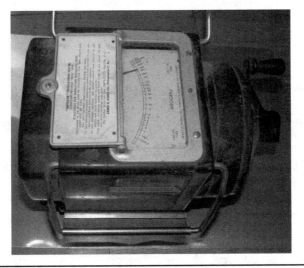

FIGURE 10-15 My old-style backup megger remains usable.

reasonably expect them to be electrical experts, nor, normally, to look at any intermediate stages of construction. All this is part of why they sometimes get me in for consultations. Third, by the same token, they can warn a customer to replace a Zinsco or FPE panel, because several electricians warned that the stabs make unreliable contact (FPE also because of the company's Listing fraud); an older ITE (and I don't just mean a Pushmatic!) because they heard the breakers might not trip; a Homeline for its termination sizes; or a fusebox as simply too old or because of AFCIs' importance. The customer often just turns around and factors this into their buy/no buy decision or their negotiation strategy.

As a contractor I've told a customer, "You ought to change this; shall I go ahead and do it?" and left it to the homeowner to decide how much work to authorize. They might suspect I'm trying to make work. As a consultant, I'm not subject to the same suspicion, but I've been asked to explain why in great detail, or to discuss degrees of risk—where there's no quantitative research on the specific issue. Then there's inspecting: my words carry a great deal more weight, as making work for a contractor doesn't earn the inspector extra money. Moreover, if need be, I can hold up the job. At the same time, there are no points to be earned for a poker face by even an inspector. Ultimately, I'm speaking to a person who I'm being paid to serve.

I'm often asked, "Is it safe?" Usually, I respond in relative terms. "This isn't to *Code*" is not that helpful when dealing with dated wiring.

Example

Rubber-insulated 10 AWG conductors in knob-and-tube wiring used to be rated at 40 amps, and 12 AWG at 25 amps. When I encounter such antique installations, there is no reason to assume that they are not grandfathered.

Still, old-doctrine wiring may be unsafe. How unsafe, compared, for example, to deteriorated insulation? Or other hazards? I'd hazard that the risk is lower, frequently, than that posed by hidden or inaccessible panelboards, even than mislabeled circuits. I can't evaluate problems that are hidden, until I hunt. Of the next seven illustrations, only Figs. 10-19 and 10-20 could be seen from the outside as problematic—and then more problems were evident once I looked inside. Without direct external evidence, and with limited time to just poke around, I have to look for the usual suspects (which set of suspects varies from situation to situation)—as imperfect a solution as I know this constitutes. But that's how I found the dangers in Figs. 10-16, 10-17, 10-18, and 10-19. I'm not shy about going after easy targets. JD Grewell poked aside a suspended ceiling tile in a basement to find the exposed luminaire insides shown in Fig. 10-21, whose one aluminum conductor, wirenutted to copper, was heavily taped where the insulation was burned off.

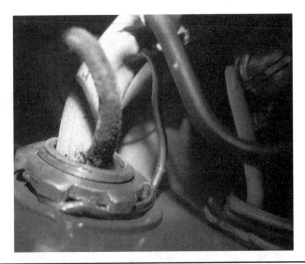

FIGURE **10-16** Not bad (if tight) for ignorant, illegal bonding.

Example

Circuits or outlets added to old systems are high in the list of "usual suspects" on consultations, and of items that warrant scrupulous inspection. Think about it: Someone added two outlets to an old structure to add dedicated circuits for window air conditioners; they even pulled a permit! And they didn't fish cable, which often would mean who knows how they ran it. They ran it nice and neat, using raceways. So how much inspection time does the job deserve? Enough time. Earlier, I showed images of UF damaged by unthinking installation.

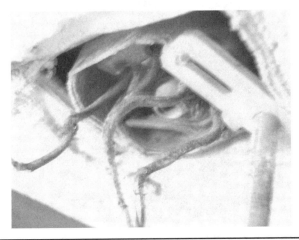

FIGURE **10-17** A canopy hid this damage.

FIGURE 10-18 Another canopy hid this damage.

FIGURE 10-19 Misapplied cable leads me to worse.

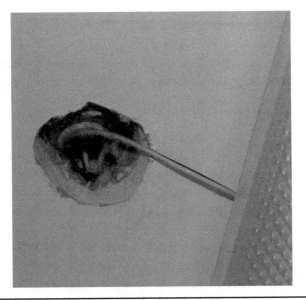

Figure 10-20 Drywall recessing the box was the first problem.

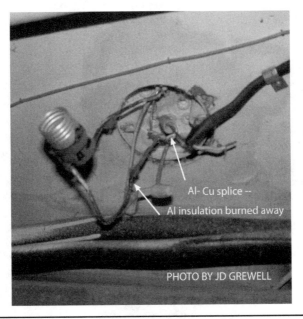

Al- Cu splice --

Al insulation burned away

PHOTO BY JD GREWELL

Figure 10-21 Lay-in tiles keep some hazards accessible.

FIGURE **10-22** Sun and heating/ cooling killed this violation.

I also see NM used in outdoor wiring; I also find plumbing fittings (Figs. 10-22 and 10-23), and raceways run without needed expansion couplings (Fig. 10-24) etc. A system that looks good, where it's easy to glance at it, can show me nastiness on the roof, or just around the corner, like the EMT in Fig. 10-25 that paid for its inadequate support.

FIGURE **10-23**   A plumbing ell led me to this enclosure.

Figure 10-24 Uncompensated contraction can cause damage.

One thing I find critical is to walk, to climb, to peer, sometimes even to crawl: not to skip equipment. An AC compressor and a bush joined forces to hide the receptacle in Fig. 10-26, but the compressor was so close I couldn't open the cover. Whatever I might say about cables and raceways, it remains true that splices and terminations, like Fig 10-27, are among the biggest trouble spots. For the overall system, grounding/bonding is so critical. A faulty water heater *will* cause electrocution, if plumbing and electrical systems are not adequately bonded. Readers have seen this.

Working with contractors whom an inspector trusts does not mean neglecting to inspect their work properly. An old-timer brought this

Figure 10-25 Unsupported, EMT bent under stress.

FIGURE 10-26   These need more than two inches to open.

FIGURE 10-27
Exposing wire in a splice is dangerously amateurish.

home to me during an inspection. I had just finished inspecting one of his jobs, and I was looking at a second one, which was very similar. In fact, both needed correction of NEC (1996) 333-7, 336-18, and 384-13 oversights. The rest was adequate on the first, and, initially, I planned to skip checking items in a particularly inconvenient location. He was surprised at this and I quickly corrected myself. I was there not only to see that he knew what he was doing, but to catch mistakes. He might have been tired or rushed on the second job.

Any structure where people sleep is a concern. Also places of assembly, which put so many people at risk. Some religious occupancies thoroughly ignore the idea of permits—or of ready access to, and clear working space around, electrical equipment. Yet coming down on community institutions can be hazardous politically. It's awkward.

CHAPTER 11

Grandfathering, Dating, and Historic Buildings

Grandfathering

Grandfathering is a topic I've touched on in several places—like the larger and somewhat-related topics of inspecting and consulting. All of these are topics that deserve their own books—or journals.

I prefer to give readers one place to go for information on how I address each topic. However, old wiring just is this way: all the aspects intermingle, and aspects of grandfathering fit naturally into several discussions. Fortunately, this is not a how-to handbook. Still, like inspection, it deserves a chapter where I can cover aspects that I didn't address elsewhere.

Dating

If something's imminently hazardous, I don't need to figure out when it was built or installed. Anything that dangerous has to go, whatever its paper trail. This is where the big guns of disconnecting utility power and withdrawing certificates of occupancy are called upon, when necessary.

Fickle Dates

Grandfathering does require knowing whether an item used to be legitimate. The first element to grandfathering, theoretically, at least, is the date a system or an item was installed. Figure 11-1 shows a transformer that clearly was a normal design—though with its wires installed through an open KO, unsecured, and, unfortunately, also unprotected by loom. This installation does not look in the least legitimate to modern eyes. The cable in Fig. 11-2, on the other hand, clearly

Figure 11-1   Legitimate!—though reidentification shouldn't look like repair.

was illegitimate from the start. Unless a practice has just been forbidden, I need to know two things: when it was legal and whether this installation took place during the period when it was legal, whether or not elements of it might have been replaced as they failed.

Three factors make the grandfathering evaluation tricky: first, many jurisdictions adopt the NEC only with local amendments; second, few adopt it on the date it is published; third and trickiest, inspectors sometimes pass work that technically is in violation—and, unfortunately, sometimes work that is profoundly in violation. I feel that way about SEC cable lying in and on the ground, unstrapped, as in Fig. 11-3, although I know some accept it. However, as unusual as the armor on the broken BX in that picture is, I don't think any inspector would have passed broken armor, however long ago—if he

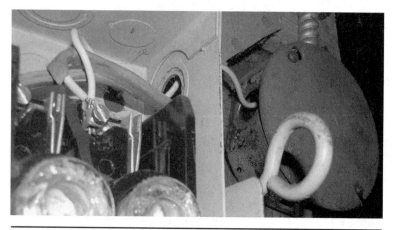

Figure 11-2   "Grandfatherable?" calls can be easy.

Figure 11-3 Broken armor is not their only violation.

noticed it. Aside from this, some designs simply are too dangerous to grandfather. Figure 11-4, for example, shows a toggle switch introduced before 1920. It would have been replaced, assuming the wiring were still functional. The conductors could conceivably still be usable. If they ran in conduit, they could have been replaced with relatively little trouble. But if it were located in a wet environment, I would want a replacement device to be suitably protected, at the least, despite the fact that the original unprotected version had been accepted.

Still, ultimately the way I know an installation warrants grandfathering, properly speaking, is that it probably started out legal and is presently still in good shape. To determine the first half of this usually requires research—research into the history of the *Code*, for one thing. It's not necessarily easy. Old *Code* books are one resource, and old *Handbooks* another (Fig. 11-5), but they're not always enough, even supplemented by ROCs, ROPs, TCRs, TCDs, and a whole assortment of other documents that didn't fit into Fig. 11-6. *Behind the Code*, by Schwan and Shapiro, is devoted to the subject of how rules came into the *Code*. It is 400-odd pages long, and full of examples. Even so, it would be quite impossible for it to serve as a reference where every rule could be looked up (and it doesn't try to).

Besides old *Code* books and *Handbooks*, sometimes I've struck paydirt by checking old magazines. That's where I found Figs. 11-7 and 11-8, the first picturing a fixture-mounting system and the second showing that hardware installed. I've dealt with these on jobs.

A distinct advance over all previous practice is represented in these switches, in which a lever movement replaces push button or key movement. The Newton Flush Toggle Switch plate is only ⅔ the size of the ordinary switch plate, has no fastening screws, is furnished in a variety of handsome styles and finishes. The Newton Surface Toggle Switch is handsome in appearance, and is self

FIGURE 11-4 Toggle switches go way back.

FIGURE 11-5 Old NECs and *Handbook*s help me date work.

FIGURE 11-6 *TCR*s and other references also help.

FIGURE 11-7 This dates a mounting hickey I have seen.

ıan sketch. this stud is ʃ... ... ...
ıt the use of bolts. The method of in-

New Kwikon "No-Bolt" Fixture Stud, Patented February 18, 1919

:tallation is extremely simple and easy

FIGURE 11-8 This one shows the hickey's intended use.

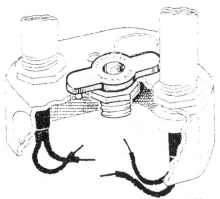

Sketch Showing Method of Installing Kwikon "No-Bolt" Fixture Stud

be enabled to finance. without drawing on their own captial. their sales of Universal Electric Clothes Washers with deferred payments running over a period of twelve months.

In seeking a plan pains were taken to work something out that would be as simple as possible and require the minimum amount of clerical work on the part of the dealer. In brief the Universal Deferred Payment Sales Plan enables the dealer to sell Universal Electric Clothes Washers to his customers— taking in payment $15 cash and executing a time payment contract under which the purchaser agrees to pay. in addition to the initial cash payment. twelve monthly installments of $14.50 each.

This contract is forwarded with the dealer's assignment to the Commercial Investment Trust who immediately remits to the dealer through his source of supply an amount which together with the initial cash payment collected by dealer is sufficient to cover the cost to him. of the machine and allow immediately. nearly one-half his profit in the transaction.

New Appleton Devices

The Appleton Electric Company. Chicago. recently brought out a line of Outlet Box Covers equipped with swivel fixture joints or hubs. for either 3/8" or 1/2" conduit: also a Unilet swivel fixture joint and a combination hickey and swivel fixture joint.

Figures No. 1 and No. 2 are raised covers showing the swivel fixture joint

These covers are far superior those formerly used and which had s

Figure No. 2

tionary hubs. The flexibility of t fixture joint or hub prevents breaki of fixture at point of suspension, a fixture will always hang plumb ev though it may be jarred through ac dent. This swivel fixture joint or h also allows for fixture to swing throu an angle of approximately 15 degr from the perpendicular, on the one st and approximately 15 degrees on other style. Covers can be furnished black enamel and galvanized finishe

Figure No. 3 is a Unilet swivel fixtu joint designed for supporting suspenc type fixture and when used in conju tion with Unilets this fixture joint s ports the fixture and the Unilet provi a separate wiring chamber for the c nection.

The use of the Unilet Swivel Fixtu Joint insures perfect allignment of

Figure No. 3   Figure No. 4

reflectors and although the fitting n be mounted on an uneven surface. ture will always hang plumb. Wl used where there are numerous fixtu in a row. this fitting will add greatl the appearance of the installation a make for greater efficiency in the lig

Figure 11-9 This dates more unusual mounting devices.

Likewise Fig. 11-9; I didn't remember coming across the swivel hubs (until I looked through my stock of takeout parts!), but it was a 1918 magazine, so while finding any of these does not tell me how new an installation is, I know that these, at least, can't be older than 1918. Figure 11-10 shows a handy BX connector; I probably have one of these in my truck at the moment, salvaged and waiting for a job on which it

would be a good match to an ancient cable. Other seemingly obvious clues to age can be deceptive. Both porcelain and Bakelite wirenuts, for example, are still available.

Even without a look-up, in extreme cases such as the use of concealed knob-and-tube from a subpanel using THWN, I would not think twice about going thumbs-down. None of my jurisdictions were still on the 1971 NEC, or had retained local amendments permitting K&T on new installations, by the time THWN was in normal use. If I'm wrong about this, or Special Permission was granted, the person whose name is on the permit needs to prove it. Other seemingly obvious clues to age can be deceptive. Listed porcelain or Bakelite wirenuts, for example, are still being manufactured.

There is another piece to the puzzle: Listing. While Listing changes and *Code* changes often progress together, this has not always been the case. However, manufacturers and NRTLs often have advised me on when Listing changes occurred. If the item isn't old, as in the broken BX in Fig. 11-3, above, or that picture's SEC and Fig. 11-11's NM run outdoors—that's not UF or SEC—and even along the ground,

Figure 11-11 NM is indoor-only—and can't just lie there.

Figure **11-12** Modified donuts were used on service entrance.

there's nothing to ponder. Also, we know that significant modification voids Listing. Unless it is plausible that the AHJ okayed a variant use, we don't have to look back to see whether an installation practice that involved modifying equipment was *Code*-compliant. Figure 11-12 shows pieced-together donuts plus a conduit bushing that did not seal the oversize opening in the back of a meter can. Despite being a noncolored image, it's clearly rusted—confirming that the space between can and wall is not a dry environment. The installer may have assumed it was dry, and assumed he could justify the non-Listed use of at least a donut. Reducing Washers are covered under UL category QCRV, which specifies, in part, "Reducing washers are intended for use with metal enclosures having a minimum thickness of 0.053 in. for nonservice conductors only." However, I can't imagine 1½ donuts being taken as acceptable.

Figures 11-13, 11-14, 11-15, and 11-16 show progressively what I saw as I began to uncover an even worse insect home. Figure 11-17 uncovers the open KO that probably was responsible for the contents— quite likely a simple, sloppy omission. The fact that the contents are disgusting isn't my point; it's hazardous, both because dry-location parts inside suffered deterioration and because a wirenut was overheating in this tinder-filled enclosure, damaging other contents. Figure 11-18, which is focused more closely at what had been the cover plate mounting holes, shows why this may have been done: a broken-off screw or screws. One legitimate alternative I tend to choose is drilling and tapping. That, to me, is within the bounds of acceptable modification, given that the support would be the same type of screw, in the original location. Figure 11-19 shows the KO closed and the

FIGURE 11-13 Tape can't replace screws

FIGURE 11-14 The tape did hold until I broke it.

FIGURE 11-15
Nonelevated/
ungasketed boxes
could easily
breach.

FIGURE 11-16 The
further I opened,
the worse it looked
inside.

FIGURE 11-17 The open KO probably was the main cause.

FIGURE 11-18 The screws froze and broke.

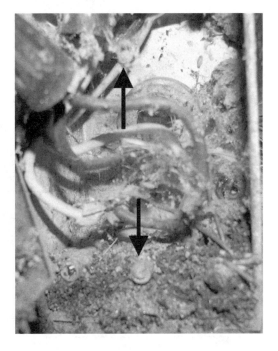

original cover, which I reused, once I'd cleared out the rubbish and restored the splices. I did substitute silicone for the missing gasket, despite this being a departure from Listed design. Figure 11-20 shows it partly restored—hole retapped—with more work to be done.

FIGURE **11-20**   Drill and tap, and I reinstall screws.

FIGURE 11-21 Duct was run up against this old pendant.

FIGURE 11-21 Duct was run up against this old pendant.

Figure 11-21 shows an old-fashioned pendant lampholder, with an HVAC duct run alongside. The light was no longer used. Figure 11-22 shows how it was made up, with a knot for strain relief and a box cover designed to support it. The other part of the old installation, the bell transformer, is in clearer view in Fig. 11-23; we see a source of concern, first broached in the discussion of grandfathering near the beginning of this chapter. Bringing wires out to the line side of the transformer through an open KO used to be standard practice, but it certainly does not come up to what modern safety doctrine tells us is appropriate. Grandfathered, yes; replaceable in kind, no.

Piecemeal Compliance

Until the mid-1970s, it was quite common for grandfathering to be determined by the "25/50% rule." If work being done over a 6-month period cost less than 25% of a building's worth (worth was calculated in various ways), the system as a whole was grandfathered with respect to building codes. Only the portion that actually was worked on had to be brought up to compliance with current codes. If work performed over the course of 6 months cost over 50% of the building's worth, the entire building needed to be brought up to date. Jobs priced

Figure 11-22 A simple knot held against a bushed cover.

Figure 11-23 RMC feeds pendant, BX, and bell power via a j-box.

in the mid-range, between 25% and 50% of total building worth, were treated in assorted ways.

Local codes are far more likely than national ones to prescribe what a contractor can and cannot do. There still is the matter of deciding what is affected by an upgrade or repair, even one required by some problem such as a plumbing leak. No jurisdiction I work in requires retrofitting AFCIs upon a service upgrade, but this could change. In Elkhart, Indiana, Chief Electrical Inspector Alan Nadon ruled that with a service change, a subpanel in a bathroom no longer was grandfathered. I respect his approach.

Incidentally, Indiana adopted the 2008 NEC statewide, with local amendments leaving out the AFCI requirement and the one for general tamper-resistant receptacles. I offer them to my customers, even though I don't work or inspect in a jurisdiction that has adopted the 2008 NEC as yet. I would not, however, urge installation of tamper-resistant receptacles on a weak, elderly person, even one with grandkids visiting, because I'm concerned about the additional hand strength needed to use them. I have known elderly people who even had difficulty inserting plugs into new standard receptacles until use had relaxed the spring metal that holds the prongs.

Even without such local interpretations, the fact that an installation is grandfathered does not mean that anything goes. Immediate, blatant hazards—for example, bare or undersized power wires—must be corrected, whatever a building's historic status.

I do find various horrors in never-inspected work. I don't include grounding to the cold water pipe, which used to be legal. ICBO (Internatonal Conference of Building Officials) rules even allowed installing a second, separate service in a residence simply because the location of the existing one was inconvenient.

I discuss unusual antiques later in this chapter. However, one type of antique could easily mislead: the box that no longer is sold, and therefore is gone from what is currently Table 314.16(a). Multiplying the "official" dimensions of an enclosure—which gives a rough-and-ready take on the volume—can mislead, because the result is definitely not the *Code*-specified minimum volume, which is the maximum that can be allowed unless the enclosure is marked otherwise. All I have to do is look at a 4 in. square by $1\frac{1}{2}$ in deep, box, what I know as the 1900: multiply the dimensions, and I get 24 in^3. Look it up, and I've lost 3 in^3—to rounded corners, to wall thickness, to whatever—compared to what simple multiplication suggested I had available. Some boxes, though, I simply don't find listed in tables of volumes. Very small pancakes such as Fig. 11-24 shows are an example of this.

If I'm replacing conductors in an odd old box, I can't make any assumption about whether it was overfilled. My modern insulation is skinnier, which may give me some slack—but the rule is tougher with respect to devices.

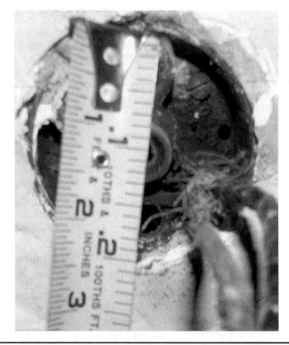

Figure 11-24   Some pancakes were much smaller than ours.

So it's nice to actually know. Here are some of the missing numbers:

In the 1968 NEC, the $1^1/2$ in deep boxes' default volumes were listed:

a $3^1/4$ in-round or octagonal box allowed 10.9 in^3;
a $3^1/2$ in-round or octagonal box allowed 11.9 in^3.

Before then, the volume tended to be stated in terms of the number of each size conductor permitted. For instance, in the 1953 edition, these were some allowances for any box shallower than $1^1/2$ in, by trade size:

3 $^1/4$ in: 4 of 14 or 12 AWG, 3 of 10 AWG.
4 in: 6 of 14 or 12 AWG, 4 of 10 AWG.

Numbers—these same numbers—had appeared by the 1940 edition. Interestingly, in leafing through various editions I found no listing for some unusual sizes, such as the 3-in round. However, this is where we first saw a specification of conductor volume, with a note that it should be used with unusual boxes. As there is no requirement back then for these to be marked with their volume, I read this as meaning the default position is indeed to multiply their dimensions, even

Figure 11-25 *Outside* clamps don't occupy box volume.

though the volumes given in 1968 showed reductions. The allowances per conductor were the ones we use today: $2\frac{1}{4}$ in per 12 AWG, etc., though the deduction of two conductors rather than one for a device is recent—it appeared in the 1990 edition, as a response to concerns about large devices such as dimmers.

> **Caution**
> The volume count includes the internal clamps, not the basically external connectors. How about external *clamps*, as in Fig. 11-25? Not counted.

Early editions, though, might offer the odd reference to sufficient room, but I find nothing specified in terms of box volume. This means that I can't be hasty in judging the legality of any original wiring in a 1920s house on the basis of crowded boxes.

However, this point goes out the window when a small older style box is already crowded with three old cables and someone squeezed in a fourth, modern one. The illegal grounding would not be necessary in order to red-tag this one, shown earlier in Fig. 9-8.

Historic Buildings

Most 1920s and even late-nineteenth-century houses are not historic buildings. Work on historic buildings adds new complexity. It raises special issues of safety, legality, value, and aesthetics—issues that are not necessarily independent. The national registry of historic sites, of official landmarks, is maintained by the National Trust for Historic Preservation. My work might affect the value of a historic house, registered or not. Decisions involving property value are my customers' choices.

Special Rulebooks

I do have to please both customers and the authorities. What this means depends on the local jurisdiction. The same rules that restrict what I can do to historic buildings occasionally also give more leeway than the NEC and the other codes adopted locally. The last time I checked, a grandfathering rule was generally adopted—through BOCA, ICC, ICBO, SBCCI, plus Californiaand several other states— and didn't require what had been installed legally to be removed, often allowing the AHJ or Fire Marshal to accept any safe alternatives to current requirements. Besides the general building code in force in each area, there are codes to address existing and sometimes specifically historic sites. Maryland has such a *Code*, relaxing rules applicable to new construction.

Interior work on historic buildings in parts of Maryland had to comply with either the NEC or the *"Smart Code"* rules, with an exemption for "ordinary maintenance or repair with no material affect on it as a historical, architectural, or cultural resource." On July 16, 2007, The ICC's *International Existing Building Code* (IEBC) was incorporated by reference as the *Maryland Building Rehabilitation Code*, with a toll-free number, 866-424-6269, for questions. The ordering page is: http://www.iccsafe.org/Store/Pages/Category.aspx?cat=ICCSafe& category=2270&parentcategory=Store%20Products&parentcategory =320.

Urgent Needs

Government-sponsored research on "common *Code* and safety deficiencies of historic buildings" included the following in its list: electrical equipment; fire alarm and detection; and lightning protection. Many historic buildings have been lost due to inadequacies in these areas.

NFPA 72, the *National Fire Alarm Code*, requires its own special study. I am not an alarm professional. Nonetheless, perhaps the most common change I have noticed being demanded by an AHJ to ready a smaller existing building for a new Certificate of Occupancy is something within my scope of expertise. This is the addition of intercommunicating, dual-source smoke alarms. In other buildings, they do require, or the owner may choose to add, a full-blown centralized alarm system.

Old, Odd, and Unsafe Equipment

Besides the more common safety defects, historic buildings present the challenges of unusual, unfamiliar equipment—some of which may be dangerous but cherished. Unfortunately, however, long it has been giving satisfactory service, the old and unusual is not necessarily safe. Portable fixtures are not my major concern. When asked to install an antique luminaire, a work of art, or an import—whether historic reproduction or avant-garde—I'm very, very leery. If the AHJ says go ahead, I'll do it. Usually they want a P.E. to sign off on the design and selection. The most reliable approach, the one I strongly prefer, is to get a field evaluation by a lab such as is associated with a NRTL. I mention UL and Met Labs to customers, because I know their field work personally, and respect it.

An ANSI standard is in process, as I write, specifying what makes a competent field evaluation, and another standard for labs doing this work. The standards, NFPA 790 and 791, probably will be final in 2011. As a third-party inspector, I certainly wouldn't give such an installation the nod without labeling such as a qualified lab can provide. Most of the time, of course, these items don't even come to the inspector's attention. This does not make installing them safe, or devoid of potential liability.

Gas to Electric

Odd fixtures are only one problem. Where the original lighting system involved gas fixtures, I have run into leftover connections to the gas system.

These can be bad news for a number of reasons. One such situation is where electrical outlets are hung from, meaning actually supported by, old gas pipes. The characteristic appearance is old black-enameled (painted) iron conduit capped not with an electrical capped bushing but with a plumbing cap, complete with sealing compound. I learned not to assume that such a pipe is empty of gas, though it probably is supplied at low pressure, if present at all. I no longer would risk dislodging such caps. If a sconce supported by such pipes is working, and the wires are good, I have had no reason to mess with it. When I'm upgrading, or the customer is too concerned to leave it be, then I move the wiring out of that box. I see no reason not to bury a pancake box, for example, once I relocate all the cables that entered it, moving them to an adjacent, larger box. I do want to leave the gas pipe accessible (unless a plumber is called in to remove it entirely).

Another hangover from gas days is the reconfigured gas light fixture. There are Listed electric post lights that are specifically designed for retrofitting gas lawn fixtures. It is perfectly okay with me to leave these in place or to install them. I just don't care to open a sealed gas line and would not use one to run wiring. When I fail to find a Listing mark on an existing, fixture that had been converted from gas to electricity, I might be inclined to consider it grandfathered, or it might strike me as too dangerous.

One troubling question remains, as with any grandfathered lumi
naire: what wattage lamp will be safe? I have been prone to suggest
using no more than a 60-W Type A incandescent lamp in any medium
base luminaire that has lost any label giving lamping instructions, but
a few years ago a UL expert told me if the label information is lacking,
it is wiser not to take a chance. If the customer insists on using the
fixture, I might mention CFLs.

Satisfying Aesthetic Requirements

Moving on from nonstandard fixtures, historic buildings present
many decisions based on the intersection of law, safety, and aesthetics. Experts in this area offer many tricks. Historic building specialist
Gersil Kay has published a lot about this work. She has managed to
hide not only wiring but also plumbing pipes and HVAC ducts in
hollow spaces of wall and ceilings. She thinks poorly of electricians
who cut into decorative moldings and sculpted surfaces for their runs,
where they can shift a few inches and cut into plain plaster, which is
far easier to restore. My key rule is to make sure the customer under-
stands what I plan, and to check with any authorities that might be
involved. Some customers opposed to wall installation of receptacles
have been open to floor outlets or pedestals; many more have not.

If I'm inspecting or evaluating the system, I want to confirm that
box and cover plate form a complete enclosure, and the contents of
each box are fully accessible. Then, I'm okay with a location that otherwise meets NEC rules, including circuiting and spacing. In mid-1997,
a house that had been wired in the 1940s still had one grandfathered
outlet going back that far. It consisted of a duplex flush receptacle with
an old-style weatherproof cover, the type having two round "doors"
that snap shut when nothing is plugged in. The receptacle it contained,
plus the cover's openings and "doors," were sized to accommodate
not modern three- or even two-prong plugs but Edison-base plugs,
like fuseholders without no-tamper inserts.

Service Changes

When I upgrade deteriorated or overloaded service equipment, I'm
getting into changes that could effect the external appearance, and
thus the historical value, of the property. So far, I have not run afoul of
the increasingly restrictive utility rules on accommodating old buildings where aesthetic issues are important. They used to be more flexible, but that's changing: safety first!

APPENDIX

Jargon and Abbreviations

T his is far from a comprehensive guide to electrical language. To find terms used in this edition that are not listed below, I offer four suggestions to those who don't have a senior colleague to consult.

First, look in the NEC's table of contents and Article 100, Definitions. Next, check the index to the NEC. After this, try a standard English dictionary. I may not be using any specialized trade definition. If you have a searchable electronic copy of the NEC, try there. Alternately, get hold of *Ferm's Fast Finder,* a very useful NEC concordance, going well beyond the NEC's index. *The American Electrician's Handbook* is another resource, with many illustrations.

Some of you may be looking for the meanings of exotic electrical terms that are not used in this volume, or for further information about particular items. In either case, older editions of *The American Electrician's Handbook* may be helpful. You can also try to get hold of the first edition of *Old Electrical Wiring,* which included a 40-page glossary of electrical terms. Finally, *Behind the Code* also has a substantial glossary, explaining numerous old terms.

1900/Four-square	A 4-in square box, most commonly $^1/_2$ in deep
8B	A 4-in diameter octagonal box, most commonly $^1/_2$ in deep
AFCI	Arc (or Arcing) Fault Circuit Interrupter
AFF	Above Finished Floor
AHJ	Authority Having Jurisdiction
AIC	Available Interrupting Current
AWG	American Wire Gauge
BX	Type AC armored cable
Classified	Tested as a compatible component for specific Listed Equipment
Cut-in Notice	A release sent from AHJ to electric utility to proceed with hookup
Dishpan; Pancake	A shallow, round box
EMI	Electromagnetic Interference

EMT/Thinwall	Electrical Metallic Tubing
EPA	Environmental Protection Agency
GC	General Contractor
GEC	Grounding Electrode Conductor
GFCI	Ground Fault Circuit Interrupter
GFPE	Ground Fault Protection for Equipment
Hickey	A threaded center support for a lighting fixture, in the context used in this book.
HO	Homeowner; often used loosely, to refer to a tenant with authority over a space.
IAEI	International Association of Electrical Inspectors
ICBO	International Conference of Building Officials, now International Code Council (ICC)
IFF	If and only If
J-box	Junction box
K&T	Knob and Tube Wiring (comments often also apply to Open Wiring)
LED	Light-Emitting Diode
Luminaire	A lighting fixture
MBJ	Main Bonding Jumper, NEC Sec. 250.28.
Mud ring	A device ring
Multi-hat	A combination inspector
NFPA	National Fire Protection Association
NM	Nonmetallic sheathed cable
NRTL	Nationally Recognized Testing Laboratory
OC	Overcurrent
OLED	Organic Light-Emitting Diode
OSHA	Occupational Safety and Health Administration
Pancake; Dishpan	A shallow, round box
Piggyback	A tandem circuit breaker
Porcelain	A unit lampholder
Postcard Permit	A permit authorizing self-certification of certain types of electrical work
PPE	Personal Protective Equipment
RMC	Rigid Metal Conduit
Service Point	The demarcation between utility and owner responsibility
SFR	Single-Family Residence
Skinny	A circuit breaker half the normal width, used in brands that did not offer tandems.
Strap	A fixture support installed across the face of, and normally supported by, an electrical box

SWD	Switching duty
TPF	Temporary Pending Final; Inspection leading to a cut-in notice
TPI	Third-Party Inspector (or Inspection) authorized to act in the AHJ's place
UL	Underwriters Laboratories, Inc.
WG	With Ground

Index